Emulsions and Oil Treating Equipment

Emulsions and Oil Treating Equipment

Editor

Sanjay Patil

scitus
academics

Emulsions and Oil Treating Equipment

Edited by **Sanjay Patil**

Printed in 2017

ISBN: 978-1-68117-360-3

Library of Congress Control Number: 2015936525

© 2016 by

SCITUS Academics LLC,
616, Corporate Way, Suite 2, 4766,
Valley Cottage, NY 10989

www.scitusacademics.com

Contents

vi

Preface

In Emulsions and Oil Treating Equipment: Selection, Sizing and troubleshooting the author provides engineers and operators with a guide to understanding emulsion theory, methods and equipment, and practical design of a treating system. This practical guide is designed to help engineers and operators develop a "feel" for selection, sizing, and troubleshooting emulsion equipment. These skills are of vital importance to ensure low operating costs and to meet crude export quality specifications. The book is written for engineers and operators, who need advanced knowledge of the numerous techniques and the equipment used to destabilize and resolve petroleum emulsions problems. Comprehensive in its scope, the author explains methods such as: demulsifiers, temperature, electrostatics and non-traditional methods of modulated or pulsed voltage control, as well as equipment such as: electrostatic treater (dehydrator), separator, gunbarr heater-treater and free water knockout.

Editor

Improving the Demulsification Process of Heavy Crude Oil Emulsion through Blending with Diluent

K. K. Salam, A. O. Alade, A. O. Arinkoola, and A. Opawale

Petroleum Engineering Unit, Department of Chemical Engineering, Ladoke Akintola University of Technology (LAUTECH), PMB 4000, Ogbomoso, Nigeria

ABSTRACT

In crude oil production from brown fields or heavy oil, there is production of water in oil emulsions which can either be controlled or avoided. This emulsion resulted in an increase in viscosity which can seriously affect the production of oil from sand phase up to flow line. Failure to separate the oil and water mixture efficiently and effectively could result in problems such as overloading of surface separation equipments, increased cost of pumping wet crude, and corrosion problems. Light hydrocarbon diluent was added in varied proportions to three emulsion samples collected from three different oil fields in Niger delta, Nigeria, to enhance the demulsification of crude oil emulsion. The viscosity, total petroleum hydrocarbon, and quality of water were evaluated. The viscosity of the three emulsions considered reduced by 38, 31, and 18%. It is deduced that the increase in diluent blended with emulsion leads to a corresponding decrease in the value of viscosity. This in turn enhanced the rate of demulsification of the samples. The basic sediment and water (BS&W) of the top dry oil reduces the trace value the three samples evaluated, and with optimum value of diluent, TPH values show that the water droplets are safe for disposal and for other field uses.

INTRODUCTION

Emulsion is defined as a system in which one liquid is relatively distributed or dispersed, in the form of droplets, in another substantially immiscible liquid. Emulsions have long been of great practical interest due to their widespread occurrence in everyday life which occurs due to reliance of the behaviour of the emulsion on the magnitude and range of the surface interaction. They may be found in important areas such as food, cosmetics, pulp and paper, biological fluids, pharmaceutical, agricultural industry, and petroleum engineering. In production and flow assurance, the two commonly encountered emulsion types are water droplet dispersed in the oil phase and termed as water-in-oil emulsion (W/O) and if the oil is the dispersed phase, it is termed oil-in-water (O/W) emulsion [1].

Water-in-oil crude oil emulsions may be encountered at all stages in the petroleum production and in processing industry. With presence of water, they are typically undesirable and can result in high pumping costs and pipeline corrosions and increase the cost of transportation [2]. Reduced throughput is needed to introduce special handling equipment, contribute to plugging of gravel pack at the sand phase [3], and affect oil spill cleanup [4].

In their research work, Micheal et al. used bottle test method to simulate field condition of four emulsion samples (two Canadian and two Venezuelan emulsions) in order to determine the variables that affect emulsion stability. They were able to evaluate response to the different emulsion based on bottle test data by introducing thirty-six different demulsifiers to enable them to probe emulsion stability. Linear regression and partition tree analysis were used to analyze the effect of various variables on emulsion stability and were able to conclude that solid content significantly affects emulsion stability. Beside solid content crude oil properties, water chemistry and process condition also influence emulsion stability [5].

Christophe et al. evaluated and compared emulsion formed by different parts of the indigenous amphiphiles (the light, the intermediate, or the heavy ones) to determine their contribution to emulsion stability. The emulsions formed with the light and intermediate fractions separated immediately when the agitation stopped. The most stable emulsions were formed with the fraction of crude that distilled at temperature greater than 520°C, suggesting that the amphiphiles with the highest molecular weight, that is, resins and asphaltenes, play a major role in the protection of water droplet against coalescence [6]. This is consistent with many recent findings that the presence of these components enhanced w/o emulsion stability (Rondón et al. and Ekott and Akpabio [7, 8]). Others factors that affect emulsion stability are fine solids, temperature, size of water droplet, and brine composition [9], which is consistent with the work of previous authors [5, 6].

Despite the success of enhanced oil recovery (EOR) process, one of the problems associated with the process is emulsion problem. Efeovbokhan et al. observed that physical factors that enhance oil recovery can also greatly contribute to the formation of very stable emulsions because EOR-induced emulsions are established by surfactant/polymer (SP) and alkaline/surfactant/polymer (ASP) processes which makes breaking of emulsion different from naturally occurring emulsions which are stabilized by asphaltenes and resins [10]. Traditional demulsifiers are often not effective on emulsions created by chemical floods; therefore, the performance of demulsifier in surfactant/polymer–flooding-induced emulsion depends on the selection of the best demulsifier with respect to the system under consideration [11]. In breaking of surfactant/polymer-flooding-induced emulsion with the use of surfactant, Oseghale et al. worked on separation of oil-water emulsions expected during chemical enhanced recovery operations using crude oil from a field in Niger delta during surfactant/polymer flooding operation. Surfactant N-octyltrimethyammonium bromide (C_8TAB) was used as the demulsifier and a dosage between 200 and

300 ppm was the optimum dose that yielded oil and water phases with oil content reduction from 550 to 70 ppm after 4 h. Microscopy test confirmed that addition of N-octyltrimethyammonium bromide (C_8TAB) produced significant coalescence shortly after it was added to the emulsion, which is in agreement with an increase of the oil droplet size in the presence of the demulsifier. Their findings show that this investigation worked with the principles of using cationic surfactants as demulsifier [12].

With various problems encountered with the presence of emulsion in our system, there is need to find ways of controlling existence of emulsion or preventing it from forming in our system. One of the ways of controlling problems encountered by crude oil emulsion is the ability to predict crude oil behaviour both at the sand phase and during production by building a robust predictive model [13]. Emulsion formation or break up either for oil in water or water in oil emulsion can be characterized based on the property and type of crude oil involved in the formation or break up of emulsion which can assist in formulating method of preventing formation of such emulsion [14]. Nuraini et al. selected four groups of demulsifiers which are amine, natural, polyhydric, and alcohol demulsifier groups serving as breaking agents of stable emulsion. Their findings show that amine demulsifier group exhibited the highest efficiency to break the emulsion compared to polyhydric, alcohol, and natural groups and that demulsifier efficiency depends on two-factor solubility of demulsifier either in water or oil and molecular weight of demulsifier [15].

It has been established from the literatures that one of the ways of breaking stable emulsion is introduction of low dose of demulsifiers. For comprehensive methods of breaking emulsion, the work of Hanapi et al. treated that aspect [2]. Micheal et al. used chemical demulsifiers and statistical analysis to classify emulsion. They obtained emulsion from the field and treated the emulsion with thirty-eight chemicals that serve as demulsifiers at nine different sites. The tests were tailored towards determination of water droplet, oil dryness, and oil-water interface which were analyzed using several statistical tools. A correlation was developed for water droplet, oil dryness and oil-water interface. The results show that water droplet significantly affect oil-water interface than oil dryness [16].

Crude oil emulsions are complex and should be characterized as completely as possible. Droplet-size distribution, interfacial phenomena, and the nature of organic and inorganic components are important. The viscosity of the emulsion is affected by both the water content and droplet size distribution [17, 18]. The increase in aqueous phase of the emulsion leads to an increase in viscosity of emulsion which in turn aggravates flow of emulsion in conduct either at the sand phase or through the surface facilities [3, 19]. Stable water-in-oil emulsions have been generally found to exhibit high interfacial viscosity and/or elasticity modulus. Viscosity of crude oil emulsion was found to increase with increase in water and decreased with increase in speed of rotation of spindle when demulsifier is added [20]. The increase of the interfacial rheological parameters has been attributed to non-Newtonian nature of emulsion [20] and physical cross-links between the asphaltene particles adsorbed at the water-oil interface [21]. Demulsification of emulsion proved to be a good method of breaking emulsions but with an influence of viscosity still unaccounted for in most of the researches; this research will study the effect of adding a diluent to emulsion samples treated with diluent for three different water in oil crude emulsions collected from three different oil fields from three operators in Niger delta, Nigeria.

MATERIALS AND METHODS

Sampling

Fresh crude oil emulsions were collected from the three oil fields flow stations operated by three different operators in the Niger delta in Nigeria, namely, Fields A, B, and C.

At the sampling points in all the three oil field mentioned above, crude oil was collected at both east and west directional sampling pipes. This to ensure that pure emulsion interface is collected and not either gas or water phase. The emulsions are collected in a tightly sealed container. The experiment was carried out after four hours from the time of sampling to avoid ageing of the crude oil. Table 1 show the initial properties of the three water-in-oil emulsion samples used for the experimental work.

Table 1: Properties of crude oil emulsions used for the analysis

	Field A	Field B	Field C
Temperature (°C)	55	60	50
Production rate (m³/day)	11,000	32,000	41000
Viscosity (mPas⁻¹)	80	100	215
Residence time (hrs)	11	7	15
API gravity (°)	23	22	21
Water cut (%)	51	8	10
Demulsifier volume used (ppm)	3	18	6

Gasoline used as the diluent for this experiment was gotten from the Nigeria National Petroleum Cooperation (NNPC) Refinery, Port Harcourt, Nigeria. Two types of demulsifiers were used by the operators for breaking of the emulsions formed. Fields A and C used PhaseTreat 4633, while Field B used PhaseTreat 6074.

The equipment used for the analysis are water bath, Checktemp1 digital thermometer, Cannon Fenske viscometer, Model HT 5001-201, six-ounce Pyrex bottles with volume 100 mL, Beaker (100 mL), Socorex Syringe micropipette, Model: Dossy TM 174 premium, Centrifuge Machine: Robinson Centrifuge, Model T.0.2, serial no. T724, Wooden product bottle shaker, and 10 mL measuring cylinder

Experimental Procedure

Each of the crude oil samples was analyzed in different setups; the three crude oil samples were treated according to the properties of the oilfield where they were collected. These properties vary in terms of temperature, rate of chemical injection, nature of process terminal, and time of processing, which will dictate the type of demulsifier chemical to be used. A water bath was set up and maintained at a temperature of 60°C equivalent to the average process temperature of the oil fields. This temperature was held constant to neglect the effect of temperature on the viscosity of the crude oil samples.

Six test bottles of capacity 100 mL were labeled according to their corresponding wells with A, B, and C, with suffixes 1 to 6 on each of the wells. The suffix 1 denotes 0 mL of diluent, 2 is 2 mL, 3 is 4 mL, 4 is 6 mL, 5 is 8 mL, and 6 is 10 mL of diluent. The bottles were filled with crude oil and gasoline to make up a volume of 100 mL. Prior to addition of diluent to the emulsion, demulsifier was added in a ratio of one-third of the amount used by the operators where the samples were collected. The samples (emulsion + gasoline) were placed in a bottle shaker and agitated thoroughly with 50 vertical shakes and 50 horizontal shakes to homogenize the diluent with the continuous phase of the emulsion. The bottle was returned to water bath after blending for ten minutes after which percentage-free water was recorded.

The viscosities reading of various combinations of the blend of demulsifier, emulsion, and diluent were obtained using the Cannon Fenske viscometer according to the procedure recommended by ASTM D445 (Norman) [22].

Basic sediment and water of the emulsion was determined using the method described in the published work of Sunil et al. [9]. Total petroleum hydrocarbon was measured by using TPH analyzer (Model HC-404). A sample of the effluent water was taken and fed into this analyzer and the reading recorded in parts per million, ppm.

RESULTS AND DISCUSSION

Flow Assurance

Effect of diluent on viscosity is illustrated in Figure 1. Gasoline was added to the three crude oil emulsions from 2 mL to 10 mL and emulsion volume is reduced from 100 to 90 mL. There is viscosity reduction when diluent is introduced. The reduction in viscosity is proportional to the increase in the volume of gasoline. Effect of viscosity reduction is a function of initial viscosity of the emulsion because from the graph the reductions of viscosity in fields A, B, and C are 38, 31, and 17%, respectively when 10 mL of gasoline was added to them and their initial viscosity are 80, 100, and 215 mPas which means that the sample with lowest value of viscosity experienced the highest percentage reduction in viscosity value and the sample with the highest viscosity experienced the lowest percentage reduction in viscosity value.

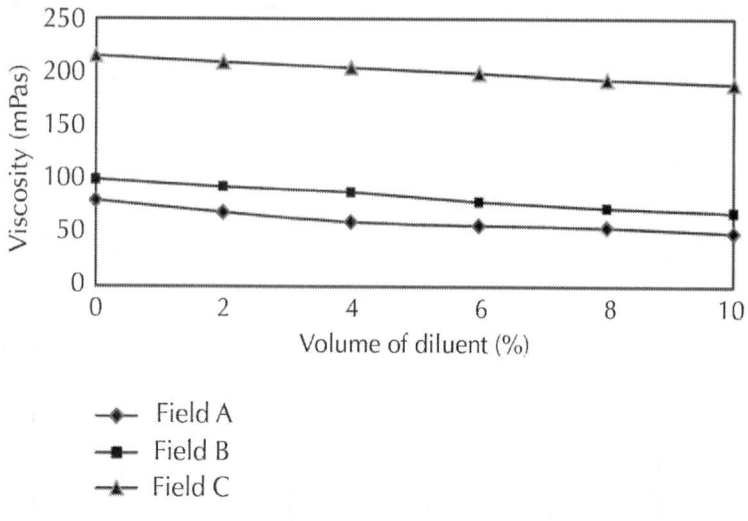

Figure 1: Volume of diluent against viscosity of emulsion.

Rate of Separation of Water

It was observed from Figures 2 and 3 that introduction of diluent affect Basic sediment and water (BS&W) of crude oil emulsion. The BS&W of Field A crude oil emulsion sample was originally 0.5% when treated with an injection rate of 1 ppm, and without blending with diluent. The value reduces as the volume of diluent increases until 8 mL when the value is zero. There is reduction in the value of BS&W of samples from fields B and C which was initially at 0.7% after the addition of 6 mL and 2 mL of demulsifiers to them to zero and 0.2% when the volume of the diluent was increased to 10 mL. Also viscosity of emulsion plays an important role in the analysis of BS&W because Field A reduces its value of BS&W to zero when the volume of diluent is 8 mL, B when the volume is 10 mL, and for C the value is at 0.2% when the volume of the diluent is at 10 mL. Field A used the lowest amount of diluent because it has the lowest viscosity while C has the the highest amount of diluent since it has the highest viscosity. Therefore, it is established that the diluent is capable of increasing reduction of BS&W in crude oil emulsions.

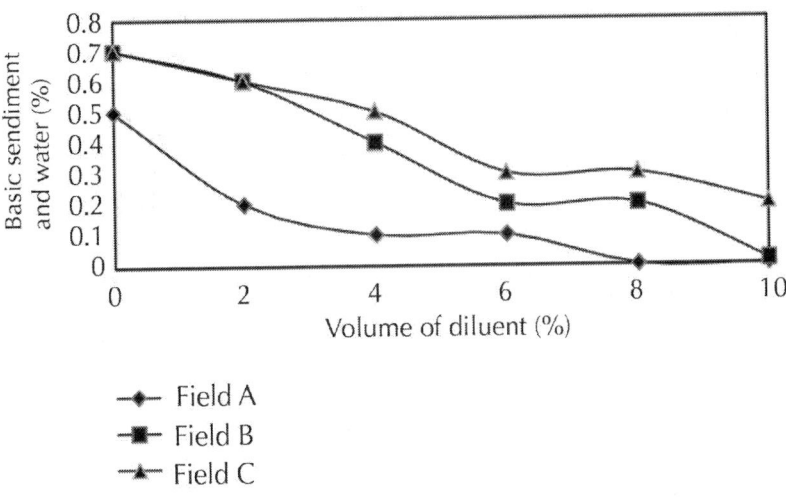

Figure 2: Volume of diluent against basic sediment and water.

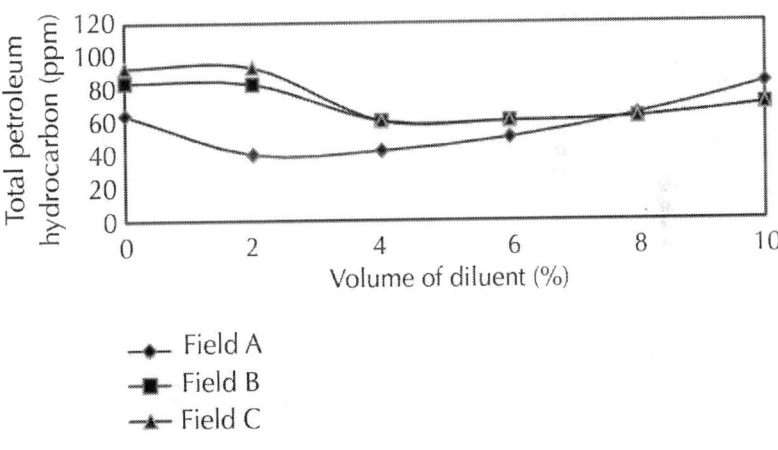

Figure 3: Volume of diluent against basic sediment and water.

Total Petroleum Hydrocarbon

The TPH of each sample initially reduces with the increase in the amount of gasoline added, but later started increasing after a particular blending point. This undesirable effect is believed to be caused by excess gasoline in the mixture finding their way into the aqueous phase.

Field A crude sample maintained a good TPH of 64 ppm which reduces as the diluent value increased between 0 and 2 mL, but there is a sharp increase in the value of TPH as the volume of diluent is increased above 2.2 mL. These reduction and rising of the TPH value with the increase in the diluent volume are attributed to the relative tightness of the crude oil emulsion. Tightness is the degree at which the water droplets are held in suspension and resist separation.

Fields B and C crude emulsions demonstrated high TPH values of 84 ppm and 93 ppm. Initially, blending showed less effect of diluent on the TPH of these two crudes between 0 and 2 mL. The TPH values of B and C reduced to 60 ppm at 4 mL of diluent, which was constant till 6 mL of diluent. Above 6 mL of diluent, there is an increase in the value of TPH fields of B and C to 70 when the volume of diluent is 10 mL. Apart from tightness of the emulsion, excess diluent can penetrate aqueous phase of the emulsion which will increase the value of its TPH.

Bottle Test

The demulsification bottle test was carried out and results on water droplet are taken after 5, 20, 30, 60, and 720 minutes. Water droplet is the separation of water from the surface of emulsion formed. The effect of addition of diluent on each crude oil sample was monitored on the rate of water droplet from each of the emulsion samples. The suffixes after the fields denotations A, B, and C indicated the variation of diluent concentration added to the emulsion samples which read 1, 2, 3, 4, 5, and 6 for 0, 2, 4, 6, 8, and 10 mL of diluent concentration.

From Figure 4, depending on the amount of diluent blended with the emulsion there was a corresponding increase in the rate of water droplet with time. When 2 mL of diluent was blended with emulsion, there was no water droplet after 5 minutes; it increased to 6% at 20 minutes and 22% at the end of 60 minutes after which there was no further droplet till the end of 720 minutes which was illustrated in A-1 in Figure 4. In A-2, the trend was similar to that of A-1 but the final water droplet value is 24%. In A-3 and A-4, there was a water droplet of 4 and 5% at 5 minutes which increased to 6 and 20% after 20 minutes, 24% at the end of 30 minutes after which there was no further droplet till the end of 720 minutes.

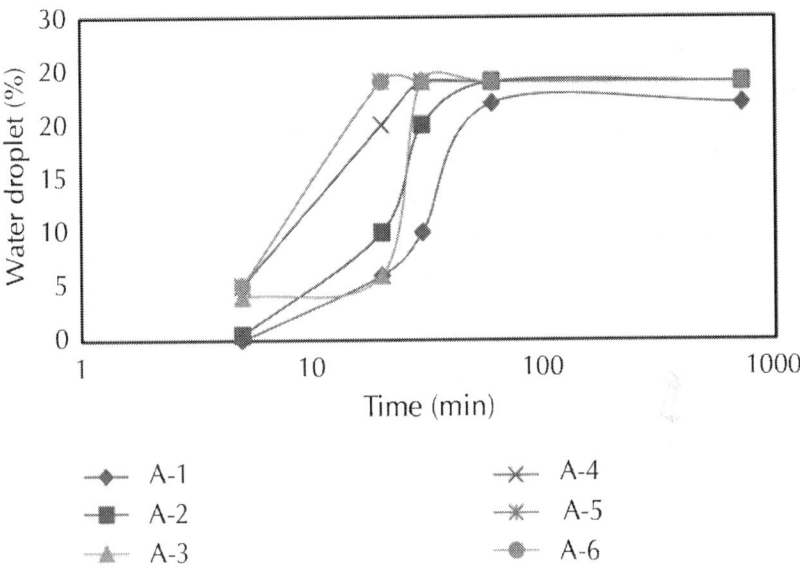

Figure 4: Water droplet against time at various emulsion/diluent ratio for Field A.

In A-5 and A-6, water droplet was 5% at 5 minutes, 24% at the end of 20 minutes, and remained constant till the end of 720 minutes. Generally, it was observed that low amount of diluent take longer time for water to drop from the emulsion but as the volume of diluent increased, the time required for water to drop out of the emulsion decreased.

Figure 5 shows the behavior of change in diluent concentration with rate of water droplet for emulsion samples collected from Field B. When no diluent was blended with the emulsion samples obtained from Field B, there was no droplet of water until after 60 minutes with a value of 4% and progressively increased to 6% at the end of 720 minutes. In B-2 to B-4, the trend of the curve followed the trend experienced in B-1 only that the rate of water droplet was faster and higher than that of B-1 with a value of 8, 10, and 10%, respectively, for B-2, B-3, and B-4 at the end of 720 minutes. In B-5 and B-6, water droplet was experienced earlier than the previous four situations with droplet of 3 and 4% at 5 minutes; it increased to 5 and 10% at 20 minutes and was 12% from 30 to 720 minutes when the analysis was terminated.

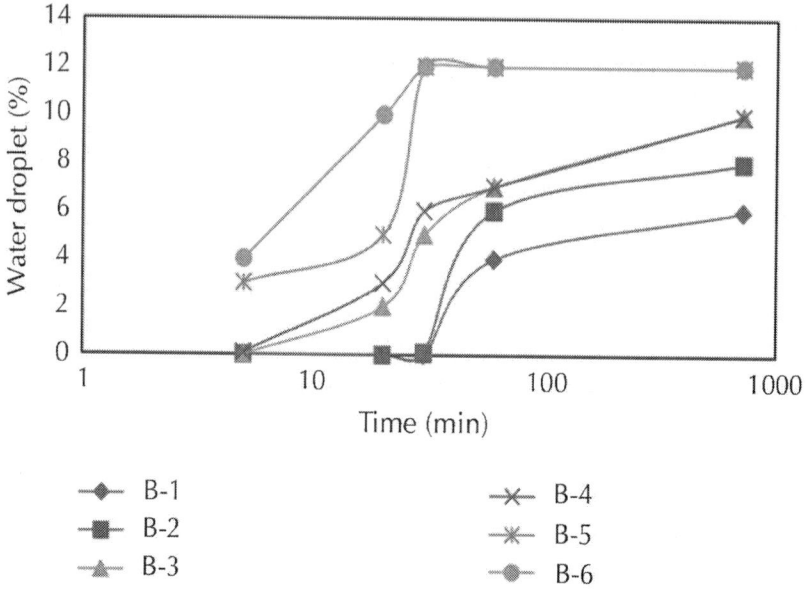

Figure 5: Water droplet against time at various emulsion/diluent ratio for Field B.

When no diluent was blended with the emulsion obtained from Field C and when 2 mL is blended with it, the behavior of their chart was similar and illustrated in C-1 and C-2 in Figure 6. There was no water droplet in the two charts until at 30 minutes with water droplet value of 0.1% which increased to 4 and 5% at 720 minutes. C-3 and C-4 followed the trend observed in C-1 and C-2 only that water droplet rate was faster with a value higher than that of C-1 and C-2. The value of water droplet at 720 minutes in C-3 and C-4 are 9 and 10%. In C-5 and B-6, water droplet was experienced earlier than the first four situation. After 5 minutes, water droplet was 1 and 2% which increased to 4 and 5% at 20 minutes. The value of water droplet remains constant at 30 till 720 minutes at 10%.

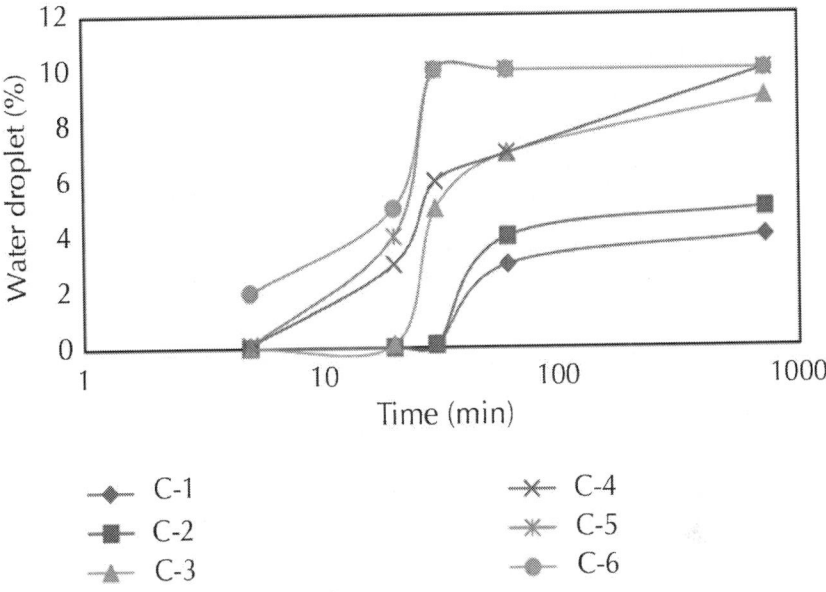

Figure 6: Water droplet against time at various emulsion/diluent ratio for Field C.

Water Quality

The quality of water droplet and observation at the oil and water interface after separation for the three crude oil emulsions are captured in Table 2. For the three crude oil emulsion samples when diluent is not blended with the emulsion, the water quality is dirty and the interface between the water droplet and oil phase is cloudy after 720 minutes. As the diluent volume blended with emulsion increased there, is an improvement in the quality of water change from dirty to clean (i.e., there is no residual emulsion or oil in the water) and the interface between oil and dropped water changes from stained to sharp (i.e., there is a distinct different between water phase and oil phase).

Table 2: Effect of diluent on water quality and interface for emulsions after 720 minutes

Volume of emulsion	Volume of diluent	Field A		Field B		Field C	
		Water quality	Interface	Water quality	Interface	Water quality	Interface
100	0	Dirty	Cloudy	Dirty	Cloudy	Dirty	Cloudy
98	2	Fair	Stained	Dirty	Stained	Dirty	Cloudy
96	4	Fair	Sharp	Fair	Stained	Dirty	Stained
94	6	Clean	Sharp	Fair	Sharp	Fair	Sharp
92	8	Clean	Sharp	Clean	Sharp	Clean	Sharp
90	10	Clean	Sharp	Clean	Sharp	Clean	Sharp

CONCLUSIONS

Generalized conclusions are hence drawn from the observation of the three samples of crude oil used for this bottle test as follows:

1. The viscosity of the three water-in-crude oil emulsions considered is inversely proportional to the increase in volume of diluent (gasoline) blended with emulsion. Also the effect of gasoline on the viscosity reduction was observed to be a function of the heaviness of the crude oil emulsion because the higher the viscosity of the emulsion the lower the reduction percentage in its viscosity value. Blending of emulsion reduced the viscosity of the three samples considered by 38, 31, and 17%, respectively.

2. Basic sediment and water (BS&W) reduces as the volume of diluent blended with the emulsion increases. This is also a function of viscosity of emulsion prior to blending because BS&W decreased with the decrease in the value of viscosity.

3. Total petroleum hydrocarbon (TPH) decreases with the increase in the volume of diluent until optimum concentration of diluent is reached and the TPH increases with further increase in volume of diluent. Optimum value varied for the three crude oil emulsions considered in the analysis. However, above optimum volume of diluent the TPH of the effluent increases which creates another problem when it comes to water disposal.

REFERENCES

1.　D. Langevin, S. Poteau, I. Hénaut, and J. F. Argillier, "Crude oil emulsion properties and their application to heavy oil transportation," Oil and Gas Science and Technology, vol. 59, no. 5, pp. 511–521, 2004.

2.　M. Hanapi, S. Ariffin, A. Aizan, and I. R. Siti, "Study on demulsifier formulation for treating malaysian crude oil emulsion," Tech. Rep., Department of Chemical Engineering, Universiti Technologi Malaysia, 2006.

3.　R. Espinoza and W. Kleinitz, "The impact of Hidden Emulsion on Oil Producing wells—stimulation concept and field result," in Proceedings of the SPE European Formation Damage, The Hague, The Netherlands, 2003, SPE paper 00082252.

4.　F. Merv and F. Ben, "Studies of the formation process of water-in-oil emulsions," Marine Pollution Bulletin, vol. 47, no. 9–12, pp. 369–396, 2003.

5.　K. P. Micheal, C. Shaokum, and C. M. Samuel, "The key to Predicting Emulsion stability: solid content," in Proceedings of the SPE International Symposium on Oil Field Chemistry, Houston, Tex, USA, 2005, SPE paper 93008.

6.　D. Christophe, A. David, S. Anne, G. Alain, and B. Patrick, "Stability of water/crude oil emulsions based on interfacial dilatational rheology," Journal of Colloid and Interface Science, vol. 297, no. 2, pp. 785–791, 2006.

7.　M. Rondón, J. C. Pereira, P. Bouriat, A. Graciaa, J. Lachaise, and J. L. Salager, "Breaking of water-in-crude-oil emulsions. 2. Influence of asphaltene concentration and diluent nature on demulsifier action,"Energy and Fuels, vol. 22, no. 2, pp. 702–707, 2008.

8.　E. J. Ekott and E. J. Akpabio, "Influence of asphaltene content on demulsifiers performance in crude oil emulsions," Journal of Engineering and Applied Sciences, vol. 6, no. 3, pp. 200–204, 2011.

9.　K. Sunil, A. Abdullah, and N. S. Meeranpillal, "An Investigative study of potential emulsion problems before field development," in Proceedings of the SPE Annual Technical Conference and Exhibition, San Antonio, Tex, USA, 2007, SPE paper 102856.

10. V. Efeovbokhan, T. Akinola, and F. Hymore, "Performance evaluation of formulated and commercially available de-emulsifiers," in Nigerian Society of Chemical Engineers Proceedings (NSChE '10), vol. 40, pp. 87–99, 2010.

11. D. T. Nguyen and N. Sadeghi, "Selection of the right demulsifier for chemical enhanced oil recovery," inInternational Symposium on Oilfield Chemistry, The Woodlands, Tex, USA, April 2011.

12. C. I. Oseghale, E. J. Akpabio, and G. Udottong, "Breaking of oil-water emulsion for the improvement of oil recovery operations in the Niger Delta Oilfields," International Journal of Engineering and Technology, vol. 2, no. 11, pp. 1–7, 2012.

13. B. Fu, "Flow assurance—a technological review of Managing fluid behaviour and solid deposition to Ensure optimum flow," in Proceedings of the 7th Annual International Forum for deepwater Technologies (Deeptec '00), Aberdeen, UK, January 2000.

14. C. Noïk, H. Malot, C. Dalmazzone, and A. Mouret, "Encapsulation of crude oil emulsions," Oil and Gas Science and Technology, vol. 59, no. 5, pp. 535–546, 2004.

15. M. Nuraini, H. N. Abdurahman, and A. M. S. Kholijah, "Effect of chemical breaking agents on water-in crude oil emulsion system," International Journal of Chemical and Environmental Engineering, vol. 2, no. 4, pp. 1–5, 2011.

16. K. P. Micheal, C. Shaokum, A. M. Robert, and C. M. Samuel, "Classifying crude oil emulsion using chemical demulsifiers and stastical analyses," in Proceedings of the SPE Annual Technical Conference and Exhibition, Denver, Colo, USA, 2003, SPE paper 84610.

17. S. D. Taylor, "Investigations into the Electrical and rheological Behaviour of W/O—emulsions in high voltage Gradients," Colloid & Surfaces, vol. 29, pp. 25–51, 1988.

18. D. G. Thompson, A. S. Taylor, and D. E. Graham, "Emulsification and demulsification related to crude oil production," Colloid & Surfaces, vol. 15, pp. 175–189, 1987.

19. T. J. Jones, E. L. Neustadter, and K. P. Whittingham, "Water-in-crude oil emulsion stability and emulsion destabilization by chemical demulsifiers," Journal of Canadian Petroleum Technology, vol. 17, no. 2, pp. 100–108, 1978.

20. N. H. Abdurahman and W. K. Mahmood, "Stability of water-in-crude oil emulsions: effect of cocamide diethanolamine (DEA) and Span 83," International Journal of Physical Sciences, vol. 7, no. 41, pp. 5585–5597, 2012.

21. J. D. McLean and P. K. Kilpatrick, "Effects of asphaltene solvency on stability of water-in-crude-oil emulsions," Journal of Colloid and Interface Science, vol. 189, no. 2, pp. 242–253, 1997.

22. J. H. Norman, Non-Technical Guide to Petroleum, Geology, Exploration, Drilling and Production, Penswell Corporation, Tulsa, Okla, USA, 2nd edition, 2001.

Synthesis and Performance Evaluation of a New Deoiling Agent for Treatment of Waste Oil-Based Drilling Fluids

Pingting Liu[1], Zhiyu Huang[1], Hao Deng[2], Rongsha Wang[2], and Shuixiang Xie[2]

[1]College of Chemistry and Chemical Engineering, Southwest Petroleum University, Chengdu 610500, China

[2]CNPC Research Institute of Safety and Environment Technology, Beijing 102206, China

ABSTRACT

Oil-based drilling fluid is used more and more in the field of oil and gas exploration. However, because of unrecyclable treating agent and hard treatment conditions, the traditional treating technologies of waste oil-based drilling fluid have some defects, such as waste of resource, bulky equipment, complex treatment processes, and low oil recovery rate. In this work, switchable deoiling agent (SDA), as a novel surfactant for treatment of waste oil-based drilling fluid, was synthesized by amine, formic acid, and formaldehyde solution. With this agent, the waste oil-based drilling fluid can be treated without complex process and expensive equipment. Furthermore, the agent used in the treatment can be recycled, which reduces waste of resource and energy. The switch performance, deoiling performance, structural characterization, and mechanisms of action are studied. The experimental results show that the oil content of the recycled oil is higher than 96% and more than 93% oil in waste oil-based drilling fluid can be recycled. The oil content of the solid residues of deoiling is less than 3%.

INTRODUCTION

Petroleum and natural gas are so important strategic resources that it is necessary to exploit them for all the countries. However, the large amount of waste drilling fluid, especially the waste oil-based drilling fluid which is produced in the process of drilling during field development [1, 2], is very harmful to the environment [3, 4]. Waste oil-based drilling fluid is classified as hazardous waste for the reason that it contains a lot of oil, heavy metals, and organic pollutants [5, 6]. Therefore, waste oil-based drilling fluid must be treated properly, or it will cause great harm to environment, animals and human [7, 8].

Because of the high oil content and stable emulsion of waste oil-based drilling fluid, treatment of it is different from that of the other drilling fluid and a great challenge [9]. If the oil in waste oil-based drilling fluid cannot be recycled, harmless processing will be very difficult to realize and the oil will be wasted [10].

Though some treatment technologies [11], such as thermal desorption, microwave processing, solvent extraction, chemical demulsification, and supercritical fluid extraction, have been used to treat the waste oil-based drilling fluid, it have been proved that all of them have some disadvantage [12–14]. Thermal desorption technology requires expensive equipment and high temperature, which causes high cost and energy consumption [15–17]. Microwave processing [18–20], which is an alternative to thermal desorption, also requires complex equipment and causes high cost. Solvent extraction technology and chemical demulsification technology require unrecyclable solvent or reagent to be added in the processing, which cause waste of resource [21, 22]. Though supercritical fluid is reusable, supercritical fluid extraction technology needs high temperature and high pressure [23, 24]. Therefore, it is very significant to develop more economic and effective methods to deal with waste oil-based drilling fluid.

In this paper, a novel surfactant for treatment of waste oil-based drilling fluid, called switchable deoiling agent (SDA), is synthesized to solve the problems of unrecyclable treating agent and complex processing condition. SDA is able to treat the waste oil-based drilling fluid without complex process and expensive equipment. Furthermore, SDA used in the treatment can be recycled, which overcomes the defect that the traditional agent can be used only once. Therefore, it is significant to simplify the processes and reduce waste of resource and energy by the method of using SDA.

MATERIALS AND METHODS

Materials

Sodium hydroxide, potassium hydroxide, sodium chloride, sodium sulfate, and hydrochloric acid were purchased from Beijing Chemical Works. Organic amine was purchased from Sinopharm Chemical Reagent Co., Ltd., China. Formaldehyde, formic acid, and tetrachloromethane were purchased from Xilong Chemical Co., Ltd., China. All the reagents mentioned above were of analytical reagent grade and used without further purification. Switchable deoiling agent for waste oil-based drilling fluid was synthesized. Waste oil-based drilling fluid was supplied by Daqing Oilfield in China.

Methods

Synthesis of SDA

A magnetic stirrer rotor was put into a two-neck bottle and kept rotating. Amine, formic acid, and formaldehyde solution were added in turns with the molar ratio of 1:5:2. When fog happens, two-neck bottle was cooled by ice packs to maintain relatively low temperature. Then, condenser tube was installed on the vertical neck of the two-neck bottle and temperature probe was installed on the other neck. The temperature was maintained at 100°C for 3 h. heating was stopped when the mixture gradually changes from light yellow to dark brown. Then, hydrochloric acid (HCl) with the same molar as amine was added when the bottle was cooled to room temperature. Heating was restarted and kept until the extra formic acid and formaldehyde were distilled out.

Then, 30% sodium hydroxide (NaOH) solution was added into the two-neck bottle to adjust the pH of the liquid to 7-8 (the color of the solution would fade from dark brown to yellow). After that, the distillation equipment was installed to distill the mixture until the liquid in the bottle turns to red. After liquid distilled out was separated, a small amount of potassium hydroxide (KOH) was added to break emulsion. The separated upper layer liquid is the target product.

Pass Gas Through

The ventilation device (as shown in Figure 1) used to pump gas into mixed liquid consists of 2 parts: a 100 mL cylinder and a 10 mm diameter aeration head connected with a 4 mm diameter flexible pipe. The gas was passed through at the speed of 1.8 L/min through the pipe and aeration head. Due to its own weight, the aeration head always remains at the bottom of the cylinder, which produces uniform and fine bubbles in the liquid and makes intensive mixture of gas and liquid.

1.8 L/min

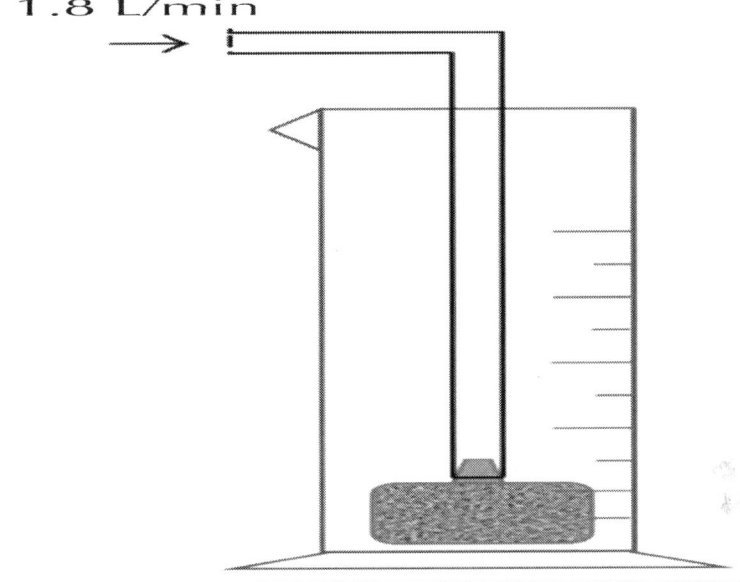

Figure 1: Diagram of ventilation device.

Determination of Oil Content in Waste Oil-Based Drilling Fluid

The water content determination apparatus, constituted by a condenser-west tube, a receiver, and the round bottomed flask, is used in this experiment. First, about 10 g (with accuracy of ±0.1 g) waste oil-based drilling fluid sample and 50 mL anhydrous petroleum ether (90–120°C) were added into the round bottomed flask. Then, the condenser-west tube and receiver were connected and the flask was heated. During the whole process, reflux rate was controlled at 2–4 drops per second. Heating was stopped when the volume of water in the receiver is no longer increased. Then, the volume of the water in the receiver was recorded. When the temperature was low enough, the rest in the flask was cleaned with anhydrous petroleum ether and filtrated with Buchner funnel. Finally, the filter residue was weighted after it was dried at the temperature of 105°C. The oil content was calculated according to the following formulas

$$H = \frac{V_w \times \rho_w}{W} \times 100\%,$$

$$S = \frac{V_f}{W} \times 100\%,$$

$$O = 1 - H - S,$$

(1)

Where H represents the rate of water content, %; O represents the rate of oil content, %; S denotes the rate of solid content, %; W denotes the weight of the sample (g); V_w is the volume of distillate (mL); ρ_w is the density of water (g/mL); W_f is the weight of filter residue (g).

Determination of Oil Content in Wastewater and Solid Waste

(A) Preprocessing. First, 10 mL acidulated water sample was mixed fully with 10 g sodium chloride (NaCl) and 20 mL tetra chloromethane (CCl_4) in a separating funnel. Then, the under-layer liquid was filtrated by a sand core funnel with 1 cm anhydrous sodium sulfate on the top. The filtrate was collected in a 50 mL volumetric flask. After that, 20 mL CCl_4 was added for the second extraction. Finally, the sand core funnel was cleaned with a small amount of CCl_4 and additional CCl_4 was added into the volumetric flask to the volume of 50 ml.

25 mL CCl_4 was added into the mixed liquid which was fully mixed with 1.00 g solid residue and 20.00 g anhydrous sodium sulfate. Then, the mixed liquid was filtrated by a sand core funnel. The filtrate was diluted to 25 mL with CCl_4 after the sand core funnel was cleaned twice with CCl_4. At last, 1 mL liquid was drawn out and diluted to 50 mL.

(B) Determination by Infrared Spectrophotometry. The oil content of liquid in 50 mL flasks was determined successively with infrared oil analyzer. The data obtained by infrared oil analyzer was multiplied by dilute multiple to obtain the oil content of oily wastewater and oily solid waste.

Structural Characterization

Fourier transform infrared spectroscopy (FTIR) (Nicolet iS50) was used for structural characterization of the deoiling agent. The samples were prepared based on pure potassium bromide (KBr) discs. First, the fully dried KBr was grinded to below 2 µm with agate mortar. After that, 70 mg grinded KBr was weighed and put into specific tableting press and then pressed to homogeneous transparent round slice under the pressure of 10 t with 5 min. Then, the pure KBr discs were impregnated into the sample solution for seconds. After that, the discs were taken out and the excessive samples were absorbed by filter paper. Finally, the prepared samples were determined with infrared spectrometer on the sample holder.

RESULTS AND DISCUSSION

Contamination in Waste Oil-Based Drilling Fluid

Five kinds of waste oil-based drilling fluid from Daqing Oilfield were used to analyze what the key components and the main pollutants were in.

As shown in Tables 1 and 2, the oil content of these waste oil-based drilling fluids is between 26.7% and 39.6%. The primary pollutant is oil. Therefore, removing and recycling the oil is the most important target for harmless treatment and resource utilization of waste oil-based drilling fluid.

Table 1: Content of each composition of waste oil-based drilling fluid

Sample source	Water content (wt. %)	Solidity content (wt. %)	Oil content (wt. %)
Unused oil-based drilling fluid (Daqing Oilfield)	58.8	14.5	26.7
Waste oil-based drilling fluid (number 501, W.S., Daqing Oilfield)	65.8	3.3	30.9
Displacing mud (number 501, WS., Daqing Oilfield)	47.3	13.1	39.6
Waste oil-based drilling fluid (number 11, X., Daqing Oilfield)	50.4	20.4	29.2

Displacing mud (number 11, X., Daqing Oilfield)	44.3	19.2	36.5

Table 2: Main pollutants in waste oil-based drilling fluid

Sample source	Cr (mg/kg)	Pb (mg/kg)	As (mg/kg)	Hg (mg/kg)	Cd (mg/kg)	Oil (mg/kg)
Unused oil-based drilling fluid (Daqing Oilfield)	18.50	14.74	14.30	1.239	0.14	267000
Waste oil-based drilling fluid (number 501, WS., Daqing Oilfield)	10.70	21.20	20.38	0.754	0.35	309000
Displacing mud (number 501, WS., Daqing Oilfield)	21.30	11.50	15.73	1.213	0.17	396000
Waste oil-based drilling fluid (numberll, X., Daqing Oilfield)	13.60	30.70	17.92	0.836	0.41	292000
Displacing mud (number 11, X., Daqing Oilfield)	9.80	16.40	10.30	0.727	0.35	365000
Control standards for pollutants in sledges from agricultural (GB4284-84)	<1000	<1000	<75	<15	<20	<3000

Performance of SDA

Switch Performance of SDA

(A) Switch Processes: The main characteristic of the deoiling agent is that its hydrophilicity can be converted. Normally, the deoiling agent is hydrophobic and stratification is obvious when it is mixed with water (Figure2 (a)). However, after gas A (CO_2) is bubbled into the mixtures, the deoiling agent becomes hydrophilic and the solution is homogeneous (Figure 2(b)). Again, the deoiling agent becomes hydrophobic and stratification is obvious after gas B (air, Ar, or N_2) is passed through (Figure 2(c)).

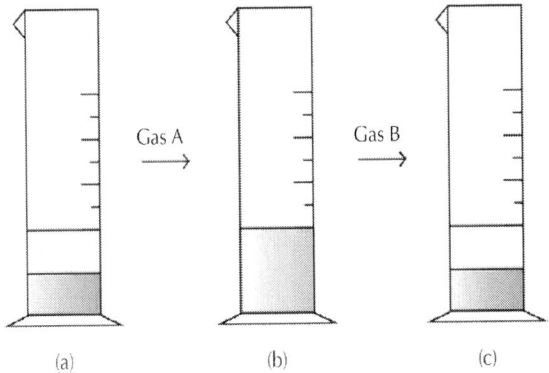

Gas A →   Gas B →

(a)   (b)   (c)

Figure 2: Switch processes.

(B) Switch from Hydrophobicity to Hydrophilicity: The experiment of switch from hydrophobicity to hydrophilicity was done under the condition of 22°C and 25% RH. First, 10 mL deoiling agent was mixed with 10 mL pure water, which formed layered liquid. Then, gas A was passed through the liquid above. Table 3 shows the time of passing gas through, volume of the hydrophilic layer, volume of the hydrophobic layer, and volume of switched deoiling agent. The conversion rate was calculated based on these results. Gas was stopped when the liquid changed to be homogenous phase, which meant that the deoiling agent was switched to be hydrophilic. In the homogenous phase, the liquid-liquid interface disappeared and the mixed liquid became slightly translucent white liquid. When deionized water was added to the uniform liquid, it also remained miscible.

Table 3: Relationship between volume change of all phases and time of passing gas through when the deoiling agent switches from hydrophobicity to hydrophilicity

Time of passing gas through (min)	Volume of hydrophilic layer (mL)	Volume of hydrophobic layer (mL)	Switched volume (mL)	Conversion rate (%)
0	10	10	0	0

2.5	12.6	7.4	2.6	26
5	15	5	5	50
7.5	17.4	2.6	7.4	74
10	20	0	10	100

Figure 3 shows the relationship between conversion rate of SDA (S) and the time (t) of passing gas A through. The blue points are experiment data and the red line is the fitting curve. The function of the fitting curve is y=x, where is conversion rate (S) and x is the time (t). That means that, when gas A is passed through at the speed of 1.8 L/min, the switch speed is 10%/min. Finally, the conversion rate rises to 100% after passing gas through for 10 min. Therefore, there is no loss in the switch process.

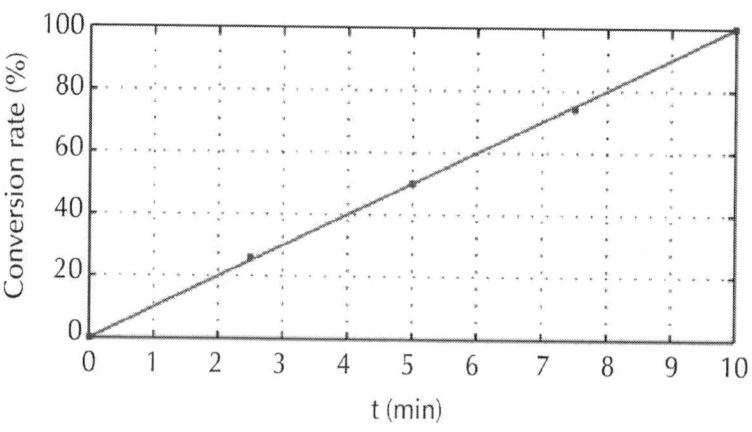

Experimental data
— y = 0.1x

Figure 3: The fitting curve of the relationship between conversion rate of SDA and the time of passing gas A through.

(C) Switch from Hydrophilicity to Hydrophobicity: The experiment of switch from hydrophilicity to hydrophobicity was done under the condition of 22°C and 25% RH. Gas B was passed into the mixed phase (prepared in Section 2.2.1) until the volume of hydrophobic layer stopped increasing. Table 4 shows the time of passing gas B through and the corresponding volume of hydrophobic layer.

Table 4: Relationship between recovery rate of SDA and time of passing gas through when the deoiling agent switches from hydrophilicity to hydrophobicity

Gas injection time (min)	Volume of hydrophobic layer (mL)	Recovery rate (%)
0	0	0
20	1	10
40	2.5	25
60	3.9	39
80	5.1	51
100	6.4	64
120	6.5	65
140	6.5	65

The highest recovery rate is 65%. Therefore, 35% SDA is lost in the process

As shown in Table 4, the volume of oil layer increases very slightly during 100–120 min and has no change during 120–140 min, which means that almost all of the deoiling agent is switched within 120 min. Figure 4shows the fitting curve of the recorded experiment data between 0 min and 140 min. The results show the relationship recovery rate of SDA and the time of passing gas B through. The function of the fitting curve is $y = 0.65x (0 < x < 100)$, $y = 65 (x \geq 100)$, where y is recovery rate (R), and x is the time (t). This means that, when gas B is passed through at the speed of 1.8 L/min, the speed of switch is 0.65%/min.

waste, %; φ_s is the oil content of diluted water extracted from waste solid determined by infrared spectrophotometry (mg/L); and ρ_w is the density of tetrachloromethane (g/mL).

Five kinds of waste oil-based drilling fluid were disposed of in the same way above; the results are shown in Table 6. As shown in this table, the oil content of wastewater is between 12.83% and 16.49%, and the average percentage is 14.51%. The oil content of oily solid waste is from 1.929% to 2.915%, and the average percentage is 2.443%. It indicates that the solid waste is harmless in China. Therefore, the treatment of the solid waste would be much easier than the treatment of the original waste oil-based drilling fluid [25, 26]. These results clearly show that SDA can significantly reduce the oil content of the oil-based drilling fluid.

Table 6: The oil content of oily wastewater and oily solid waste

Number	The measured value of water (mg/L)	The measured value of solid (mg/L)	Oil content of oily wastewater (mg/L)	Oil content of oily solid waste (%)
1	73.525	33.384	14.71	2.616
2	82.437	24.617	16.49	1.929
3	66.553	37.194	13.31	2.915
4	75.968	28.752	15.19	2.253
5	64.142	31.931	12.83	2.502
Average	—	—	14.51	2.443

Structural Characterization and Mechanism of Action

Structural Characterization

Fourier transform infrared spectroscopy was used to analyze the structure of composite product (Figure 6). The location of the absorption peaks and corresponding transmittance were found after the data from spectrogram were processed.

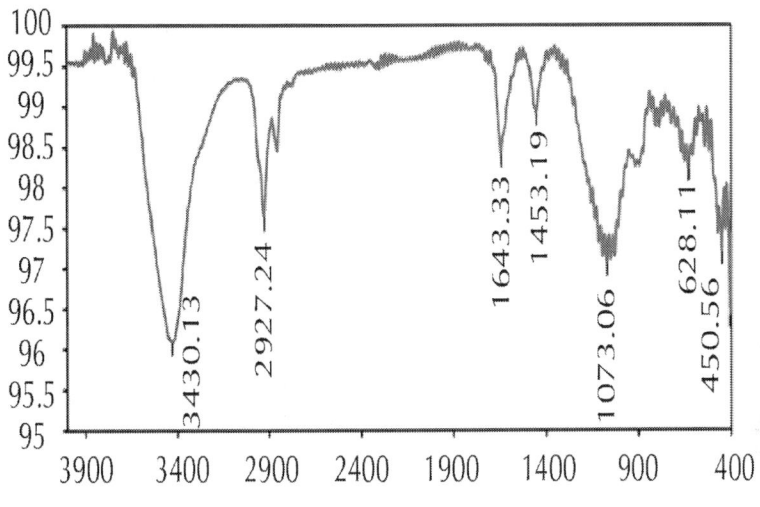

(a)

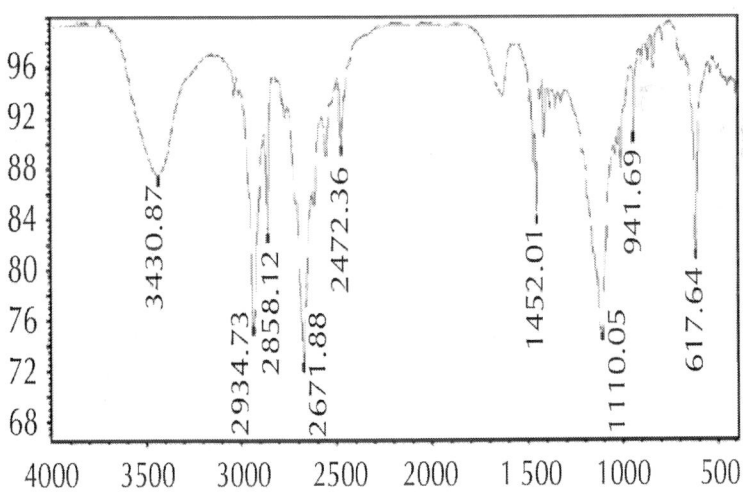

(b)

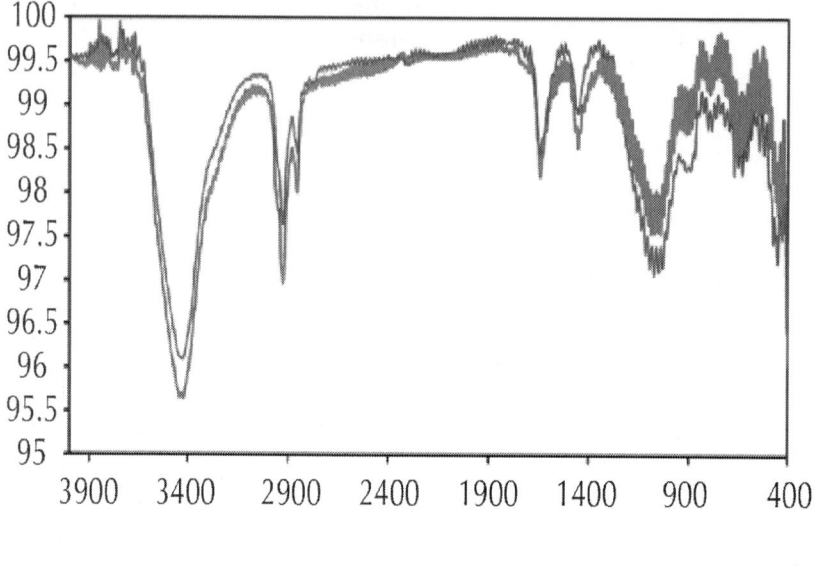

(c)

Figure 6: FTIR spectra of SDA.

Figure 6 shows the infrared spectroscopies of switchable deoiling agent in three stages of the switching progress. Figure 6(a) shows the infrared spectroscopy of original SDA, which is hydrophobic; Figure 6(b) shows the infrared spectroscopy of hydrophilic SDA switched from original SDA; Figure 6(c) shows the comparison of the infrared spectroscopy of the original SDA and the infrared spectroscopy of hydrophobic SDA switched back from hydrophilic SDA.

In Figure 6(a), the adsorption peak at 1073.06 cm^{-1} has relationship with C–C bond. The two peaks at 1349 cm^{-1} and 1453.19 cm^{-1} come from C–H bonds in methyl. Another peak at 2927.24 cm^{-1} is observed as methyl, methylene, and methane. The emergence of peak at 3430.13 cm^{-1} means the existence of secondary amines or tertiary amines. According to the analysis results, the synthetic product contains large amounts of methyl, amine, and secondary carbon atoms. So it is speculated that synthetic product is based on one or several amines as the main component.

Comparison of Figures 6(a) and 6(b) shows that the position of absorption peak in Figure 6(a) is similar to the position of absorption peak is Figure 6(b). The main differences between these two figures are as follows: (1) in Figure 6(b), there is an obvious absorption peak at 2671.88 cm^{-1}, which means the emergence of ammonium cations; (2) in Figure 6(b), the absorption peak is pronounced weaken at 3430.87 cm^{-1}, which means the reduction of tertiary amine functional group. Therefore, the main change in the progress of switching hydrophobic SDA to hydrophilic SDA is that the tertiary amine is converted to ammonium salt with the effect of CO_2 (gas A) and water.

In Figure 6(c), the red curve is the infrared spectroscopy of the original SDA which is similar to the infrared spectroscopy in Figure 6(a). The blue curve is the infrared spectroscopy of the recycled SDA switched from hydrophilic SDA by passing gas B through. The blue curve shows high similarity to the red curve, which means that the recycled SDA has little difference with the original SDA and can be reused.

Mechanisms of Action

It is the main function for switchable deoiling agent, which has the characteristic of controllable hydrophilic transformation, to remove and recycle oil in the waste oil-based drilling fluid. Its main action principle is composed of deoiling mechanism and transformation mechanism.

The deoiling mechanism is based on the principle of the dissolution in the similar material structure. It refers to the similar structure and intersolubility of the solute and solvent, which means that in this paper the polar solutes dissolve in polar solvents while nonpolar solutes dissolve in nonpolar solvents. Waste oil-based drilling fluid is emulsion composed of water, oil, and solid impurities and so on. For the reason that the oil phase is mainly nonpolar and the water is polar, the oil phase is easily extracted by nonpolar solvents. And, in general, deoiling agent, which is hydrophobic, can absorb oil to realize oil-water separation. However, when gas A is passed into the mixture, oil is separated alone because deoiling agent has integrated with water for its change of hydrophilic nature. Similarly, after gas B is passed into the aqueous phase, deoiling agent is separated from water because it is hydrophobic again. So, the deoiling agent can be reused:

$$NR_3 + H_2O \underset{+\text{gas B } (-\text{gas A})}{\overset{+\text{gas A}}{\rightleftharpoons}} R_3NH^+ + (OH^- + \text{gas A})$$

(5)

The switching mechanism is that hydrophobic amine, the main composition of deoiling agent, can react reversibly according to the change of gas. As shown in the conversion mechanism (5), the hydrophobic amines (R is saturated alkyl) react with gas A to form hydrophilic product when gas A is passed into the aqueous phase. However, the reaction reverses to form original hydrophobic amine when gas B is passed into to replace gas A. Because of different reaction rate and activation energy the forward and converse reaction needs, there is great different conversion rate between the two reactions.

CONCLUSIONS

In summary, hydrophilicity of deoiling agent used for waste oil-based drilling fluid is convertible according to the need of human. Normally, deoiling agent is hydrophobic. However, it is hydrophilic when gas A (CO_2) is passed into and it is hydrophobic again when gas B (air, Ar, or N_2) is passed into. It is effective to use deoiling agent to deal with waste oil-based drilling fluid. The oil removal rate can reach 94% and the oil content of extracted oil is about 97%. The residues of deoiled waste oil-based drilling fluid are water phase and solid phase. The test result shows that the oil content of wastewater is below 17 mg/L and the oil content of oily solid waste is below 3%.

According to the analysis results of FTIR, the synthetic product contains large amounts of methyl, amine, and secondary carbon atoms. The amines especially are the leading parts. The deoiling mechanism of the switchable deoiling agent is based on the principle of the dissolution in the similar material structure. The oil is extracted from the waste oil-based drilling fluid based on the hydrophobicity of deoiling agent and the deoiling agent is recyclable for its switchable performance. The transition mechanism of it is mainly based on the reversible reaction among the amines with the water and gas. Thus, using SDA to treat the oil-based drilling waste fluid can recycle not only the oil in the fluid but also the SDA itself, which reduces waste of resource. Furthermore, with SDA, the waste oil-based drilling fluid can be treated without complex process, expensive equipment, and harsh conditions. Therefore, this technology can significantly reduce the waste of resources, energy, and the cost, comparing with the commonly used technologies.

The conversion process of deoiling agent is rapid from hydrophobic to hydrophilic and the oil loss during this process is negligible. However, the recovery process of deoiling agent, switch from hydrophilic to hydrophobic, is slow and the oil loss cannot be neglected in this process. Therefore, the next important research is how to improve the conversion rate from hydrophilic to hydrophobic and reduce the oil loss from hydrophilic to hydrophobic.

ACKNOWLEDGMENTS

The authors gratefully acknowledge the technical support of Laboratory of Solid Waste Disposal and Comprehensive Utilization of CNPC and the financial support provided by National Science and Technology Major Projects of China (2011ZX05021-004).

REFERENCES

1.   S. A. Mahmoud and M. M. Dardir, "Synthesis and evaluation of a new cationic surfactant for oil-well drilling fluid," Journal of Surfactants and Detergents, vol. 14, no. 1, pp. 123–130, 2011.

2.   A. A. Hafiz and M. I. Abdou, "Synthesis and evaluation of polytriethanolamine monooleates for oil-based muds," Journal of Surfactants and Detergents, vol. 6, no. 3, pp. 243–251, 2003.

3.   S. Rana, "Facts and data on environmental risks-oil and gas drilling operations," in Proceedings of the SPE Asia Pacific Oil and Gas Conference and Exhibition, SPE112993, Perth, Australia, October 2008.

4.   R. Sadiq, T. Husain, B. Veitch, and N. Bose, "Risk-based decision-making for drilling waste discharges using a fuzzy synthetic evaluation technique," Ocean Engineering, vol. 31, no. 16, pp. 1929–1953, 2004. · ·

5.   T. Steliga, P. Jakubowicz, and P. Kapusta, "Changes in toxicity during in situ bioremediation of weathered drill wastes

contaminated with petroleum hydrocarbons," Bioresource Technology, vol. 125, pp. 1–10, 2012. · ·

6.   Kisic, S. Mesic, F. Basic et al., "The effect of drilling fluids and crude oil on some chemical characteristics of soil and crops," Geoderma, vol. 149, no. 3-4, pp. 209–216, 2009. · ·

7.   P. J. Cranford, D. C. Gordon Jr., K. Lee, S. L. Armsworthy, and G. H. Tremblay, "Chronic toxicity and physical disturbance effects of water- and oil-based drilling fluids and some major constituents on adult sea scallops (Placopecten magellanicus)," Marine Environmental Research, vol. 48, no. 3, pp. 225–256, 1999. · ·

8.   M. Denoyelle, F. J. Jorissen, D. Martin, F. Galgani, and J. Miné, "Comparison of benthic foraminifera and macrofaunal indicators of the impact of oil-based drill mud disposal," Marine Pollution Bulletin, vol. 60, no. 11, pp. 2007–2021, 2010. · ·

9.   R. S. Farinato, H. Masias, D. Garcia, R. Bingham, and G. Antle, "Separation and recycling of used oil-based drilling fluids," in Proceedings of the International Petroleum Technology Conference (IPTC ‹09), IPTC13238, Doha, Qatar, December 2009.

10.   R. Sørheim, C. E. Amundsen, R. Kristiansen, and J. E. Paulsen, "Oily drill cuttings—from waste to resource," in Proceedings of the SPE International Conference on Health, Safety and Environment in Oil and Gas Exploration and Production, SPE61273, Stavanger, Norway, June 2000.

11.   M. Charles and S. Sayle, "Offshore drill cuttings treatment technology evaluation," in Proceedings of the SPE International Conference on Health, Safety and Environment in Oil and Gas Exploration and Production, SPE126333, Rio de Janeiro, Brazil, April 2010.

12.   T. Hagan, L. R. Murray, T. Meling et al., "Engineering and operational issues associated with commingled drill cuttings and produced water re-injection schemes," in Proceedings of the SPE International Conference on Health, Safety and Environment in Oil and Gas Exploration and Production, SPE73918, Kuala Lumpur, Malaysia, March 2002.

13.   E. Malachosky, B. E. Shannon, J. E. Jackson, and W. G. Aubert, "Offshore disposal of oil-based drilling-fluid waste:

an environmentally acceptable solution," SPE Drilling and Completion, vol. 8, no. 4, pp. 283–287, 1993.

14. H. Shang, C. E. Snape, S. W. Kingman, and J. P. Robinson, "Microwave treatment of oil-contaminated North Sea drill cuttings in a high power multimode cavity," Separation and Purification Technology, vol. 49, no. 1, pp. 84–90, 2006. · ·

15. J. T. Araruna Jr., V. L. O. Portes, A. P. L. Soares et al., "Oil spills debris clean up by thermal desorption,"Journal of Hazardous Materials, vol. 110, no. 1–3, pp. 161–171, 2004. · ·

16. R. L. Stephenson, S. Seaton, R. McCharen, E. Hernandez, and R. B. Pair, "Thermal desorption of oil from oil-based drilling fluids cuttings: processes and technologies," in Proceedings of the SPE Asia Pacific Oil and Gas Conference and Exhibition (APOGCE ‹04), SPE88486, Perth, Australia, October 2004.

17. A. J. Murray, M. Kapila, G. Ferrari, D. Degouy, B. J. Espagne, and P. Handgraaf, "Friction-based thermal desorption technology: Kashagan development project meets environmental compliance in drill-cuttings treatment and disposal," in Proceedings of the SPE Annual Technical Conference and Exhibition (ATCE ‹08), SPE116169, Denver, Colo, USA, September 2008.

18. M. S. Pereira, C. M. de Ávila Panisset, A. L. Martins, C. H. M. de Sá, M. A. de Souza Barrozoa, and C. H. Ataíde, "Microwave treatment of drilled cuttings contaminated by synthetic drilling fluid," Separation and Purification Technology, vol. 124, pp. 68–73, 2014. ·

19. J. P. Robinson, S. W. Kingman, C. E. Snape et al., "Scale-up and design of a continuous microwave treatment system for the processing of oil-contaminated drill cuttings," Chemical Engineering Research and Design, vol. 88, no. 2, pp. 146–154, 2010. · ·

20. J. P. Robinson, S. W. Kingman, C. E. Snape et al., "Remediation of oil-contaminated drill cuttings using continuous microwave heating," Chemical Engineering Journal, vol. 152, no. 2-3, pp. 458–463, 2009. · ·

21. Z. Talbi, B. Haddou, Z. Bouberka, and Z. Derriche, "Simultaneous elimination of dissolved and dispersed pollutants from cutting oil wastes using two aqueous phase extraction methods," Journal of Hazardous Materials, vol. 163, no. 2-3, pp. 748–755, 2009. · ·

22. A. A. Abdel-Azim, A. M. Abdul-Raheim, R. K. Kamel, and M. E. Abdel-Raouf, "Demulsifier systems applied to breakdown petroleum sludge," Journal of Petroleum Science and Engineering, vol. 78, no. 2, pp. 364–370, 2011. · ·

23. R. B. Eldridge, "Oil contaminant removal from drill cuttings by supercritical extraction," Industrial and Engineering Chemistry Research, vol. 35, no. 6, pp. 1901–1905, 1996.

24. C. G. Street and S. E. Guigard, "Treatment of oil-based drilling waste using supercritical carbon dioxide,"Journal of Canadian Petroleum Technology, vol. 48, no. 6, pp. 26–29, 2009.

25. C.-H. Chaineau, J.-C. Setier, and A. Morillon, "Is bioremediation a solution for the treatment of oily waste?" in Proceedings of the 10th Abu Dhabi International Petroleum Exhibition and Conference, SPE78548, Abu Dhabi, UAE, 2002.

26. R. K. Dhir, L. J. Csetenyi, T. D. Dyer, and G. W. Smith, "Cleaned oil-drill cuttings for use as filler in bituminous mixtures," Construction and Building Materials, vol. 24, no. 3, pp. 322–325, 2010.

Chapter 3

Current Trends in Water-in-Diesel Emulsion as a Fuel

Mohammed Yahaya Khan[1], Z. A. Abdul Karim[1], Ftwi Yohaness Hagos[1], A. Rashid A. Aziz[1], and Isa M. Tan[2]

[1]Department of Mechanical Engineering, Universiti Teknologi Petronas, Seri Iskandar, 31750 Tronoh, Perak, Malaysia
[2]Department of Fundamental & Applied Science, Universiti Teknologi Petronas, Seri Iskandar, 31750 Tronoh, Perak, Malaysia

ABSTRACT

Water-in-diesel emulsion (WiDE) is an alternative fuel for CI engines that can be employed with the existing engine setup with no additional engine retrofitting. It has benefits of simultaneous reduction of both NO_x and particulate matters in addition to its impact in the combustion efficiency improvement, although this needs further investigation. This review paper addresses the type of emulsion, the microexplosion phenomenon, emulsion stability and physiochemical improvement, and effect of water content on the combustion and emissions of WiDE fuel. The review also covers the recent experimental methodologies used in the investigation of WiDE for both transport and stationary engine applications. In this review, the fuel injection pump and spray nozzle arrangement has been found to be the most critical components as far as the secondary atomization is concerned and further investigation of the effect of these components in the microexplosion of the emulsion is suggested to be center of focus.

INTRODUCTION

Diesel engines offer better fuel to power conversion efficiency and due to their better fuel economy, diesel engines are the dominant class of engines in mass transportation, heavy industries, and agricultural sectors. In spite of their preferable advantages, they are one of the major pollution contributors to the environment. Primary pollutants emitted from diesel engines are particulate matters (PM), black smoke, nitrogen oxides (NO_x), sulphur oxides (SO_x), unburned hydrocarbon (HC), carbon monoxide (CO), and carbon dioxide (CO_2) [1]. Increasing stringent regulation on exhaust emissions drives a major research endeavour in engine development in order to reduce these pollutants [2, 3]. Significant reduction targets include reduction of PM from 0.025 g/km in Euro 4 (2005) to 0.0045 g/km in Euro 6 (2014) for both CI passenger cars and light commercial vehicles, which account for a 82% reduction. Similar reduction targets are also imposed on heavy-duty engines with a reduction of 50% in PM emission [4].

Modern hardware-based solutions for pollution control such as diesel particulate filters (DPF), high-pressure fuel injection equipment (FIE), and sophisticated piezoinjectors and associated control systems are avenues being followed by engine designers and manufacturers. However, these technologies come with high price tags and cannot be fitted to existing engines. Therefore, there is a pressing need for appropriate technology that can be applied to these existing engines. One such possibility is to develop fuel-based solutions which do not rely on new hardware to control the combustion process and hence the emissions. Research showed that WiDE used as an alternative fuel in CI engines can lead to reductions in the adiabatic flame temperature resulting in measurable reductions in the NO_x emissions [5–8]. There are many advantages to using emulsion fuels, such as more complete combustion, leading to better fuel economy, and cleaner burning fuels with fewer emissions.

The main mechanism causing the reduction in NO_x emissions seems to be the decrease in the temperature of the combustion products as a result of vaporisation of the liquid water and consequent dilution of the gas phase species. As for PM emissions, the presence of water during the intensive formation of soot particles seems to reduce the rate of formation of soot particles and enhance their burnout by increased concentration of oxidation species such as OH [9].

Water can be introduced into the combustion chamber in different ways as follows: (a) introduction of water with the inlet air in liquid or vapour form, (b) parallel water and diesel injections, and (c) WiDE with or without surfactants. While the first two methods of water introduction are subjected to additional cost of water injection system and engine corrosion problems [10], the later method has been regarded as the most effective technique for the simultaneous reduction of both particulate matters and NO_x [11]. Moreover, WiDE is a convenient renewable fuel option as the existing engine does not require any prior or postmodification.

Till date, research on WiDE is active and even its comparative advantage to its base petroleum fuels is not precisely known. The reasons behind this is the lack of understanding of the combustion phenomenon associated with the formation of soot inside the combustion chamber, complexity in nature of combustion chamber, unknown end-to-end physical path of emulsion (evaporation and mixing), and the effect of microexplosion phenomenon inside the combustion chamber. Because of these reasons, the reported result of researches in this area has been inconsistent [9] as far as the brake thermal efficiency, brake specific fuel consumption, and pollutant formation are concerned. In addition to the above mentioned problems, there are many factors affecting the atomization and the general combustion process on top of the wide operating variables. The researches were mainly focused in specific engine operation variables and due to this, the results has become very difficult to draw a general conclusion. The results reported by different researchers are often conflicting, sometimes generating results that are even worse than pure diesel fuel [12]. As a result, there is still a need for further investigation especially with fuelling WiDE into compression ignition engine by varying the applied conditions. This paper will review the current status of water-in-diesel emulsions so that to bring research works in this area under one document and to further enlighten the possible area of intervention for researchers.

METHODOLOGIES USED IN WATER-IN-DIESEL EMULSION STUDY

Different methodologies have been used to study WiDE as a fuel for internal combustion engine both inside and outside the engine combustion chamber. Abu-Zaid [13] has used horizontal stainless steel and aluminum surfaces to study the evaporation of water-in-diesel and water-in-kerosene emulsion by varying the surface temperature from 100 to 460°C at atmospheric pressure. The evaporation characteristics of the droplet, effect of water concentration, and total evaporation time were investigated experimentally. Tanaka et al. [14, 15] and Tsue et al. [16] used the same horizontal hot surface to study the microexplosion of evaporating droplet. While the hot surface used by Abu-Zaid [13] was exposed to the atmosphere, the apparatus used by Tanaka and coworkers [14, 15] and Tsue and coworkers [16] was made of duralumin and the hot surface was isolated from the atmosphere with a high pressure cylindrical chamber. The major objectives of the experimental investigations made by this apparatus were to study the effect of ambient pressure on the start of microexplosion [15], the effect of the water concentration, base fuel property, and surface temperature on the statistical characteristics of the start of microexplosion [14], and to study the start of microexplosion of the emulsion fuel droplet by the use of statistical analysis [16]. Watanabe et al. have used single droplet experiment where the emulsified droplet was suspended on a fine wire to study the breakup characteristics of a secondary atomization of an emulsion [17], Jeong et al. [18] studied the autoignition and microexplosion behavior of a single droplet, Jeong and Lee [19] investigated the autoignition and microexplosion behaviors of one-dimensional arrays of fuel droplets, and Morozumi and Saito [20] examined the microexplosion characteristics of the emulsion droplet. Yatsufusa et al. [21] have used air-assisted fuel atomizing burner to study the combustion and emission characteristics of WiDE.

The application of both hot surface and suspension single droplet as a means to examine both the evaporation and microexplosion phenomena is very important to predict the air-fuel mixing process and further the combustion and emission formation process. The results might not be accurate as the apparatuses are constructed with major assumptions. However, experimentation of evaporation and microexplosion phenomena and investigation of the effect of these phenomena in the combustion and emission formation inside the combustion chamber are extremely difficult tasks.

There have been other practices and methodologies used in the study of WiDE other than the above mentioned ones, such as WiDE and other test fuels in a diesel like constant volume combustion chamber and rapid compression expansion machine with controllable temperatures and pressures in the range of 293–923 K and 0.1–5.0 MPa, respectively [12, 20–26]. The effect of temperature and pressure on the microexplosion phenomenon was experimentally investigated by using this apparatus with the help of a multipulsed ruby laser holocamera with an off-axis image-plane optical path and a high speed camera [12]. The effects of injection pressure and water concentration on the spray combustion characteristics, like ignition delay and lift-off length of combustion of the emulsion, were investigated with the help of a diesel-like constant volume chamber. The high pressure and high temperatures were created by combustion of carbon-monoxide mixed with compressed air and oxygen and ignited by a spark plug [27]. Water-in-diesel microemulsion, WiDE, and conventional diesel fuels were experimentally investigated for their physical properties, spray behavior, and combustion characteristics. Spray cone angle, liquid phase penetration, droplet penetration, and vapor penetration were studied prior to ignition inside a controlled atmosphere similar to engine combustion chamber in a high pressure high temperature test rig. In this study, the size of water droplets in the WiDE and water-in-diesel microemulsion was further investigated with the help of a Nuclear Magnetic Resonance (NMR) diffusometry. Physical characterization of the test fuels was also achieved with controlled stress rheometer for rheology, sigma 70 tensiometer for surface tension, and Mettler Toledo DA-100 M density meter for density measurements [22]. The effect of water concentration and injection timing on the combustion performance and emission of an emulsified fuel were experimentally investigated in an engine like rapid compression and expansion machine, or RCEM. Successive flame image by high speed camera, pressure data, needle lift, and crank angle were taken in every 0.1° crank angle intervals [26].

Different types of four stroke engines with and without an optical access have been used to study the effect of WiDE on combustion and emissions [7, 28–35]. All researchers used similar arrangement in this type of experimental investigation except for the variation of size and technology of the components. The setup consisted of four stroke engine connected to eddy current dynamometer, high pressure-transducer for engine cylinder indicating to be instrumented in one cylinder, fuel flow rate meter, thermocouples for engine inlet and exhaust emissions, air inlet flow rate measurement, and exhaust gas analyzer for the measurement of NO_x, CO, CO_2, HC, and O_2. Study of the combustion process was difficult since there has not been any optical access to the combustion chamber. As a result, more focus was given to the cylinder and thermocouple readings for the engine wall and inlet and exhaust temperatures to study the performance of combustion.

There are also significant researches of WiDE for CI engine application by numerical and mathematical modeling. These studies can be broadly categorized into modeling water-in-diesel spray with a focus on the microexplosion, droplet break-up and autoignition [12, 36–40]; heat release and engine performance modeling of combustion [41, 42] and emission formation modeling [43, 44].

PRINCIPLE OF WATER IN DIESEL EMULSION

An emulsion is a mixture of two or more liquids immiscible in nature, one present as droplet, or dispersed phase distributed throughout the other, or the continuous phase [54]. It is generated by means of a mechanical agitation in the presence of surface active agents, sometimes called emulsifiers or surfactants, for stability. The surfactants possess a polar, or hydrophilic head and a nonpolar, or hydrophobic tail [55]. It is incorporated to weaken the surface tension of the medium in which it dissolves. When it is placed in an oil-water mixture, the polar groups orient towards the water and the nonpolar group orients towards the oil as it lowers the interfacial tension between the oil and water phases [56]. They are classified into cationic, anionic, amphoteric, and nonionic based on the type of polar group on the surfactant. For a best formation of appropriate surfactant, hydrophilic-lipophilic balance, or HLB (water liking-oil liking) score is developed. Low HLB tends to make water-in-oil-emulsion while those with a high HLB are more hydrophilic and tend to make oil-in-water-emulsion. The value of HLB ranges from 1 to 20.

The fact that emulsion is used as a fuel in diesel engine, it is recommended that it should be stable and this can be realized with the help of suitable surfactants. Surfactants should easily burn with no soot and free of sulfur and nitrogen [36]. Furthermore, they should have no impact on the physiochemical properties of the fuel. Usually the amount introduced in the emulsion process is in the range of 0.5–5% by volume ratio. The most common surfactants used in the water-in-diesel emulsion are sorbitan monooleate [9, 17, 30, 32] and polyethylene glycol sorbitan monooleate mixture [13, 24, 57, 58], polyethylene glycol sorbitan monooleate (polysorbate 80) and sorbitol sesquioleate (SSO) mixture [22], sorbitan monolaurate [27], gemini [32], polyoxyethylene nonylphenyl ether [15, 59, 60] solgen 40 and noigen TDS-30 (dai-ichi kogyo seiyaku) [16], polysorbate 20 (commercially known as tween 20) [33], detergent/liquid soap [21, 61], and t-octylphenoxy polyethoxy ethanol.

There have been a limited literature about the effect of surfactant on the characteristics of water-in-diesel emulsion as far as combustion and emissions are concerned [5]. Nadeem and coworkers have studied water-in-diesel emulsion with conventional (sorbitan monooleate) and gemini surfactants for main pollutant emissions by fuelling it in a four stroke and four cylinder engine test bed and concluded that for 15% water content, there is 71% reduction in PM emission with gemini surfactant water in diesel emulsion fuel [32].

There are two types of emulsification techniques, namely, two-phase (sometimes called primary) and three-phase emulsion (sometimes called multiphase or secondary emulsion to include complex emulsions with more than three liquid ingredients). The two-phase emulsion constitutes one continuous phase and one-dispersed phase liquids while the three-phase emulsion constitutes one continuous phase and two or more dispersed phase liquids. Although this paper is mainly focused at the water-in-diesel emulsion which is categorized under the two-phase emulsion, addressing the three-phase emulsion research, specifically those which are compared with two-phase emulsions is essential. Therefore, research work in the three-phase emulsion techniques has also been reviewed.

Three-Phase Emulsion

Two types of three-phase emulsions can be resulted from the three-phase emulsification technique (Figure 1) depending on the inner and outer phases, namely, oil-in-water-in-oil and water-in-oil-in-water emulsions. Oil-in-water-in-oil emulsions are applicable for fueling purposes in the internal combustion engine while water-in-oil-in-water emulsion is applied in cosmetics, food, or pharmaceutical manufacturing [1]. A limited literature is available on the oil-in-water-in-oil emulsions application as a fuel in internal combustion engines. Three-phase emulsion can be prepared by three techniques, namely, phase inversion, mechanical agitation, and two-stage emulsion [1]. A two stage emulsification technique has been used for the preparation of three-phase oil-in-water-in-oil emulsion by many researchers [30, 31, 42, 43]. This technique, which is the most common technique in three-phase emulsion, uses both lipophilic and hydrophilic type of surfactants. First, a two-phase oil-in-water emulsion is prepared by using a hydrophilic type surfactant and a mechanical homogenizer machine. A lipophilic type of surfactant is then used to further emulsify the two-phase oil-in-water emulsion in oil and form three-phase oil-in-water-in-oil emulsion.

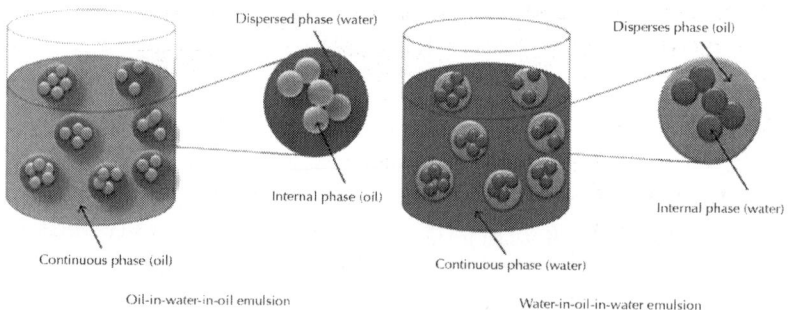

Figure 1: Concept of three-phase oil-in-water-in-oil and water-in-oil-in-water emulsions.

On study of emulsification characteristics property of three-phase oil-in-water-in-oil, Lin and Wang [1] have investigated the effect of homogenizing machine speed, oil/water ratio, HLB, and surfactant amount on the diameter of the liquid droplets, viscosity, and general stability of the three-phase emulsion. Based on their conclusion, there was a decrease in the diameter from $6\mu m$ to 2-3 μm of the liquid droplets with an increase in the stirring speed of 2500 rpm to 7500 rpm. The viscosity of the three-phase oil-in-water-in-oil emulsion was also highly influenced by the oil/water ratio in which viscosity increased with an increase in water content in the inner phase. Furthermore, more stable three-phase emulsion was reported with a surfactant volume of 2% and HLB value in the range of 6–8 [1]. Another study by the same authors [31] was conducted to determine the effect of HLB value, water content, oil/water ratio, stir speed, and engine operating conditions on the performance and emissions of four-cylinder four stroke marine diesel engine. They have compared the two-phase and three-phase emulsion and the general emulsions with the base diesel fuel based on the engine performance and emission parameters. They have reported that the three-phase emulsion has lower brake specific fuel consumption, CO, NO_x, and equivalence ratio but higher exhaust gas temperature compared to the two-phase emulsion. Moreover, the emulsion registered a lower exhaust gas temperature, O_2, NO_x, and smoke opacity, and higher CO_2, CO, and equivalence ratio compared to the base diesel fuel. Similar research by Lin and Chen [30] was conducted to investigate the effect of emulsifying mechanism on the performance and emission of two-phase and three-phase emulsion fueled in four-cylinder four stroke marine diesel engine. In this study, they have compared both two-phase water-in-oil emulsion and three-phase oil-in-water-in-oil emulsions prepared with ultrasonic vibrator and mechanical homogenizer. The same authors, on a separate publication, have also investigated the effect of the time of emulsification, quantity, and HLB of surfactant on the emulsified fuel properties of two-phase water-in-oil and oil-in-water emulsions and three-phase oil-in-water-in-oil emulsion prepared by ultrasonic vibrator. They have compared oil-in-water and water-in-oil two-phase emulsions with respect to their temperature rise with increase in emulsification time, emulsion stability, and the size of dispersed phase. They have reported lower temperature rise with an increase in emulsification time, evenly distributed and smaller size of dispersed phase, and highest stability with oil-in-water emulsion [58].

With the fact that the three-phase oil-in-water-in-oil emulsion uses two-stage emulsification processes and requiring two types of surfactants, generally the process cost would be greater than that of a two-phase emulsion process. Although the above mentioned literature documented the advantages of three-phase emulsions over the two-phase emulsions, it is not yet clear where the boundary lies as far as the process cost, emulsion characteristics, emulsified fuel properties, engine performance, and emission are concerned. It is also not clear about the comparison of the micro explosion process of two-phase and three-phase emulsion. A further study on the micro explosion phenomenon of three-phase oil-in-water-in-oil emulsion fuel inside a combustion chamber atmosphere is recommended.

Two-Phase Emulsion

There are two basic forms of two-phase emulsion. The first is the oil-in-water (O/W) emulsion in which oil droplets are dispersed and encapsulated within the water column. The second is the water-in-oil (W/O) emulsion in which droplets of water are dispersed and encapsulated within the oil. Figure 2 shows the concept of two-phase water-in-oil and oil-in-water emulsions. For either type of stable emulsion to form, three basic conditions must be met [62].

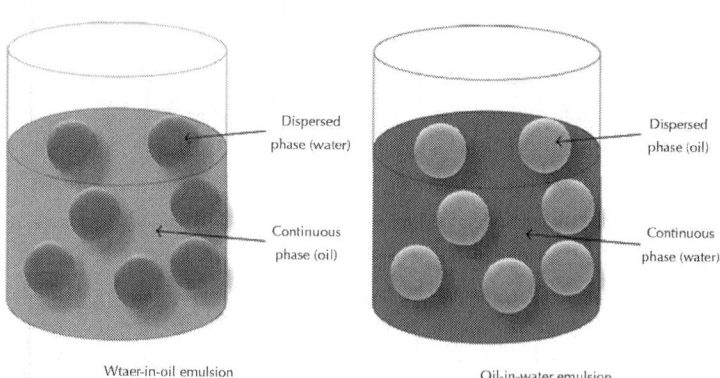

Wtaer-in-oil emulsion   Oil-in-water emulsion

Figure 2: Concept of two-phase water-in-oil and oil-in- water emulsions.

(a) The two liquids must be immiscible or mutually insoluble in each other. (b) Sufficient agitation must be applied to disperse one liquid into the other. (c) An emulsifying agent (surfactant) or a combination of emulsifiers must be present. In addition to the above emulsion, few researches included concepts of introducing three-phase emulsions and comparative studies on the effect of two-phase and three-phase on the diesel engine performance also available. Irrespective of method of production either by mechanical homogenizing or by ultrasonic vibrating, the oil-in-water-in-oil (O/W/O) emulsions were found to have a higher fuel consumption rate, brake specific fuel consumption, CO emission, and black smoke opacity than the W/O emulsions [30, 31]. Surfactants used for the formation of water-in-diesel emulsion fuel should burn easily with no soot and should be free of sulphur and nitrogen as discussed in [5]. Furthermore, they should have no impact on the physicochemical properties of the fuel. Surfactants from the aliphatic hydrocarbon family are the best candidates to be used as emulsifiers. Usually the amount of surfactants used for emulsification is in the range of 0.5–5% by volume, as the surfactant concentration increased emulsion stability reduced.

EFFECT OF EMULSIFICATION ON STABILITY AND PHYSIOCHEMICAL PROPERTIES OF WIDE

The stability of the diesel emulsion is affected mainly by the emulsification technique, emulsification duration, volume fraction of water (dispersed phase), viscosity of continuous phase (diesel oil), stirring speed (or ultrasonic frequency), and concentration of surfactants. The experimental work by Chen and Tao [62] studied the effect of emulsifier dosage, oil water ratio, stirring speed, and emulsifying temperature on the stability of water in diesel emulsion using mechanical agitator. They reported that an increase in oil to water ratio, stirring speed, and duration had positive influence on stability, whereas an increase in emulsifying temperature showed negative impact. Two-phase W/O emulsion showed better engine performance with less CO emissions which were reported in [30] with the application of ultrasonic vibrator compared to the emulsion prepared by mechanical agitation. In addition, the selections of suitable surfactants, the choice of a suitable agitator frequency, and agitation time have also been identified as equally important parameters in the formation of stable emulsified fuels [63].

Surfactant or emulsifier is the most important factor that affects the stability of an emulsion. Percentage of water in the emulsion, stirring intensity, stirring duration, emulsifying temperature, and operational pressure also affect the stability of an emulsion. Chen and Tao [62] have experimentally studied the effect of emulsifier dosage, oil-water ratio, stirring speed, and time and emulsifying temperature on the stability of diesel-water emulsion. They have concluded that an emulsifier dosage of 0.5%, oil-water ratio of 1 : 1 by volume, stirring speed of 2500 rpm, and duration of 15 min and emulsifying temperature of 30°C have been optimum for the stability of the emulsion. They have also reported that while an increase in oil to water ratio, steering speed, and duration up to 15 min have positive impact on stability, increase in emulsifying temperature had negative impact. Similar work has been done by Ghannam and Selim on the stability of water/diesel emulsion fuel and they indicated the necessity of surfactant for the stability of the emulsion and possibility of getting stable emulsion of higher water percentage (>30%) by increasing the percentage of emulsifier agent (2%) and increasing speed up to 20000 rpm with mixing period of 30 min [64].

To be a good fuel for a compression ignition engine, a water-in-diesel emulsion should possess most of the positive effects of petrodiesel fuel. As this type of engine is well established, complete alteration of fuel characteristics that requires engine retrofitting would not be feasible economically. A good CI engine fuel should hold characteristic features like short ignition lag, sufficiently high cetane rating in order to avoid knocking, suitably volatile in the operating range temperatures for good mixing and combustion, easy startup characteristics, limited smoke and odor, suitable viscosity for the fueling system, free from corrosion and wear, and ease of handling [65]. In diesel engines, fueling system must insure the fuel to be delivered into the engine cylinder economically and in an appropriate time so that it runs smoothly with minimal exhaust and noise. This is done by controlling the process of spray penetration, chemical and physical atomization, mixture ignition, and combustion and exhaust gas formation. These are mainly dependent on the physiochemical behaviors of the fuel and the injection system. Significant number of the literature could be found on the physiochemical behavior of water-in-diesel emulsion as well as their effect on the combustion behavior and stability. As the water content of the emulsion increases, physical properties like density [7, 32, 66], viscosity [7, 26], bulk modulus of elasticity [7], and compressibility [44] increased. A very close attention should have to be given to these changes as density has a pronounced effect on the mixing process and viscosity in the injection system. It has also been reported that water addition reduces the heating value of the emulsion.

EFFECT OF WIDE ON COMBUSTION PROCESS

In a WiDE, water remains embedded inside the diesel droplets with the help of the surfactants. When this type of emulsion is sprayed into a hot combustion chamber, heat is transferred to the surface of the fuel droplets by convection and radiation. Since water and diesel have different boiling temperatures, the evaporation rates of these two liquids will be different. As a result, the water molecules reach their superheated stage faster than the diesel, creating vapor expansions breakup [17, 20, 37, 40, 58, 67]. It is at this stage that the two phenomena, microexplosion and puffing, prevail. Microexplosion is that the whole droplet breaks up into small droplets quickly, while in puffing, water leaves the droplets in a very fine mist (a part of the droplet break up) [22, 26].

These microexplosions result in a fast breakdown, or secondary atomisation, of the fuel droplets which in turn causes fast fuel evaporation and hence an improved air-fuel mixing as illustrated in Figure 3. Therefore, it is equally important to study the basics of microexplosion of water-in-diesel emulsion and its influencing parameters as it plays a major role in the combustion improvement.

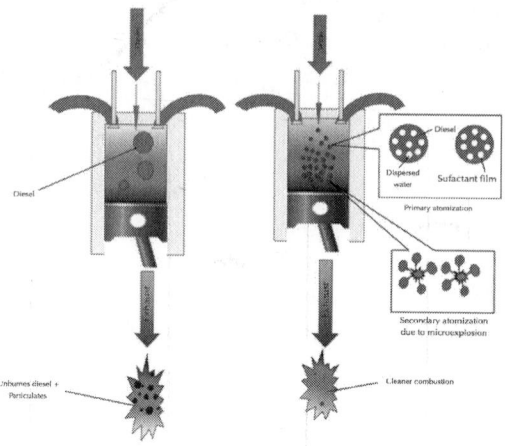

Figure 3: Primary and secondary atomisation in spray flame of emulsified fuel.

According to Morozumi and Saito, micro explosion is mainly affected by the volatility of the base fuel, type of emulsion, and water content. Based on their conclusion, an increase in an emulsifier content increases the micro explosion temperature and waiting time [20]. Mechanisms of microexplosion and their dependence on various parameters affecting microexplosions have been extensively investigated by Fu and coworkers [40]. They have stated that both water-in-oil and oil-in-water emulsions can microexplode at certain conditions. Furthermore, they have related the diameter of the dispersed liquid with the strength of the microexplosion through a physical model.

The advantages on performance and emissions of water-in-diesel fuel and factors affecting microexplosion have been extensively researched. Fu [68] and coworkers have challenged its occurrence inside diesel engine combustion chamber. Based on their conclusion, the droplet diameter of an emulsion in the combustion chamber is in the range of 20–30 µm and microexplosion phenomenon could not occur with this range of droplet sizes [38]. Even though this report is agreeable with the reports that state the effect of mean water particle size diameter on the intensity of microexplosion [39, 69], it contradicts with most of the literature on the occurrence of microexplosion in diesel engine combustion chamber.

Microexplosion is an important phenomenon in the secondary atomization process of water-in-diesel emulsion fuels. Generally, this phenomenon is affected by volatility of base fuel, type of emulsion, water content, diameter of the dispersed liquid, location of the dispersed liquid, and ambient conditions like pressure and temperature. Although many studies have been conducted both experimentally and numerically to understand the phenomenon of microexplosion, yet the study of its effects inside the combustion chamber are quite few. It is believed that fuel injection and the passage of emulsified fuel through the narrow exit of the injection nozzle affect the dispersed liquid behaviour of the fuel. It is therefore very important to study the microexplosion phenomenon inside a combustion chamber and its effect on the combustion process like the secondary atomization, spray penetration, evaporation, and mixture ignition.

Combustion process is generally characterized by factors such as injection characteristics, spray penetration, evaporation, chemical and physical atomization and mixture ignition, engine cylinder pressure and temperature, and heat release characteristics [8, 22, 41, 70]. As far as the fuel-injection characteristics is concerned, it is observed that the injection pressure profile extension over a longer period leads to retarded injection timing and 22–26% increase in injection duration [41]. Armas et al. have also reported similar results on injection and they have associated it with an increase in viscosity of the emulsified fuel [7]. Ochoterena et al. have studied the spray behaviors of WiDE, water-in-diesel microemulsion and conventional diesel on high pressure and high temperature constant volume chamber, keeping an eye on the penetration and lift-off, cone angle measurements, start of combustion measurements, and singularities of atomization. A longer droplet penetration and wider cone angle with the emulsified fuel compared to pure diesel fuel was observed which was associated with a lower volatility of water [22]. A slightly longer ignition delay, same report by Ghojel and Honnery [41] and Armas et al. [7], and longer combustion duration were also reported with the emulsion fuel, both as a result of lower flame temperature. Ignition delay up to 29% was reported when WiDE as fuel in an HSDI diesel engine [45]. In another experiment Subramanian et al. [46] reports that the ignition delay is much higher with WiDE as compared to water injection to the manifold during the intake stroke. The effect of water content, the injection pressure, and ambient temperature on ignition delay was further studied by Ghojel and Tran [27]. Ambient temperature significantly affected the ignition delay. On the other hand, no significant effect was observed with injection pressure. They have also studied the effect of water content, the injection pressure, and ambient temperature on flame lift-off. They have reported an increase in flame lift-off with an increase in injection pressure and water content while it decreased with an increase in ambient temperature. Alam Fahd et al. [11] has experimentally found that pressure traces and heat release rate were comparable with respect to pure diesel at different speed and loading conditions.

EFFECT OF WIDE ON ENGINE PERFORMANCE

The effect of volume percentage of water added in the emulsion on the performance of an engine was studied by many researchers [26, 32, 33, 44, 47–50, 70]. Abu-Zaid has studied torque, power, brake specific fuel consumption, and brake thermal efficiency by varying the volume percentage of water from 0 to 20% water/diesel ratio with 5% resolution [47]. Alahmer et al. studied the above mentioned engine performance parameters in a four stroke, four-cylinder direct injection engine by varying the volume percentage of water from 0 to 30% water/ diesel ratio with 5% resolution [33]. The experimental investigation by Selim and Elfeky [70] on the other hand has used 0, 2, 4, 6, and 8% by volume of water in the emulsion to study its effect on the heat flux on the engine components. Water contents of 5%, 10%, and 15% by volume were used in the study of their effect on the engine performance parameters (torque, power, brake mean effective pressure, and specific fuel consumption) [32]. Here, surfactant type was also taken as a variant to see the effect of gemini surfactant on the engine performance parameters and its comparison with conventional ones. Park et al. have experimentally studied the effect of the volume percentage of water on the combustion characteristics of an emulsion fuel in a rapid compression and expansion machine by considering 0, 16.67%, and 28.6% of water by volume in the emulsion [26]. In another study, Park et al. have experimentally investigated the combustion characteristics and engine durability of a four-stroke, six cylinder direct injection diesel engine with a turbocharger used as a power unit in city/highway bus fueled by pure diesel, 13%, 15%, and 17% of water by volume in the emulsion [48]. Kannan and Udayakumar have also experimentally studied the effect of water percentage of water emulsified diesel fuel on the brake thermal efficiency, brake specific fuel consumption, NO_x, and hydrocarbon emissions in a single cylinder four stroke direct injection diesel engine by considering 0, 10%, and 20% of water by volume in the emulsion [44]. On a separate study by Samec et al., water contents of 0, 10%, and 15% by volume were considered for the experimental investigation of the effects of water content on the combustion characteristics of diesel engine [49, 50]. This experimental work was accompanied with a numerical investigation.

As far as the combustion performance and emission (refer to Table 1) of diesel engines fueled with WiDE are concerned, inconsistent results have been reported by different researchers. Besides, all the reports are based on different engine setups and methodologies. As a result, an optimum percentage of water content in the emulsion cannot be drawn. But we can conclude from this that water content ranging from 5 to 40% by volume in the emulsion can be utilized for fueling transport and stationary diesel engines. A systematic approach of studying the optimization of water content in the emulsion for best engine performance and emission by both experimental and numerical investigations is necessary that can give best recommendation for the commercialization of WiDE as an alternative source of energy of the future diesel engines.

Table 1: Engine Performance for WiDE under various testing conditions

Reference	Engine type and loading conditions	% of water	Surfactant used	Amount of surfactant used	% increase in specific fuel consumption	% increase of torque	% increase of Brake thermal efficiency	% reduction NO_x	% reduction of PM	% reduction of HC and CO
[7]	Renault F8Q turbo charged intercooler IDI,5 different steady state operating conditions	10	Polyethylenglycole monooleate and sorbitol sesquioleate	NA			Reduced	Reduced	Reduced	HC reduced
[8]	4s, 4c, di industrial diesel engine	NA	NA	NA	22–26% compared with certified diesel fuel (CDF)	NA	Slightly Higher than CDF	29–37% reduced	Not measured	60–90% reduced HC
[11]	2.5 L DI turbo-charge Toyota diesel engine, 25%, 50%, 75%, and 100% load with 800–3600 rpm in steps of 400 rpm	10% water	10% biodegradable surfactant	10% by volume	Increased in all test conditions	NA	Increased with speed	Reduced	NA	Higher at low load and decreasing with increasing speed and load
[32]	FORD XLD 418, 1000–5000 rpm	5, 10 and 15	Conventional-sorbitan monooleate (SM) and gemini surfactant	0.5% for SM 0.4% for gemini	15% water has highest and decreases with decrease in water content	Less with all emulsions compared to diesel 5% water produced highest torque	NA	Reduced	Reduced	Lowest with 15% water

Ref	Engine	Water content	Surfactant	Surfactant amount	Increased (BSFC/Performance)	At 5% water torque was max., and declining with increase of water content	For 5% water = 35%	NO and NO$_x$ reduced with increase in water	Reduced	HC and CO increased with increases in water content
[33]	4 cylinder, DI water cooled 1450cc, 1000–3000 RPM	5–30% in steps of 5	Polysorbate-20	2% by volume	Increased	At 5% water torque was max., and declining with increase of water content	For 5% water = 35%	NO and NO$_x$ reduced with increase in water	Reduced	HC and CO increased with increases in water content
[41]	4C,4S, water cooled DI industrial HINO diesel engine, 200 Nm and 2200 rpm	13	NA	2% (surfactants and cetane improver)	Increase of 26%	NA	NA	NA	NA	NA
[44]	Single cylinder, 4S, DI diesel engine with injection pressure of 200 bar, constant speed 1500 rpm	10% and 20%	Sodium laurel sulphate	0.1% for 1000 mL emulsion	Break BFC decreases with all load	NA	Increase with increased water content	Reduced 10% for 10% water and 25% for 20% water	NA	Decreasing with all loading conditions
[45]	4 cylinder, HSDI diesel engine at 1480, 2035, 1480, 2065, and 1460 rpm	20	Span 80 and Tween 85	1.3% of Span 80 and 0.7% of Tween 85	BSFC increased with increased EGR rate	NA	NA	Reduced between 30–50% at low injection pressure and increased up to 24% at higher injection pressure	94% reduced at low loads	
[46]	4S, air cooled overhead valve, constant speed of 1500 rpm at different outputs.	0.4 : 1 ratio	Surfactant used unknown, with HLB 7	NA	NA	NA	NA	NO$_x$ is reduced	NA	NA
[47]	Single cylinder DI diesel engine, 1200–3300 rpm	0–20% in steps of 5%	Span 80 and tween 80	2% by volume of mixture	Decreased with increasing water content	NA	Appr-3.5% for 20% water	NA	NA	NA

Ref	Engine	Content	Additive								
[48]	6 cylinder, TCI diesel engine (High way Bus Engine), 10, 25, 50, 75 and 100% of full power at 1200 rpm and 2000 rpm	15% analysis considered for 15% water content	NA (used along with cetane improver)	NA	at 25% and 50% load slightly higher than diesel. At 75% BSFC is better than diesel	Decreased 20% and 9% at 1200 rpm and 2000 rpm when compared with diesel	NA	Reduced up to 11.6%	Reduced up to 34.5%	CO and HC increased up to 12.4% and 59.4% respectively
[49]	4 cylinder air-cooled DI truck diesel engine	0, 10, 15	(Span 85) Quantity NA	NA	NA	NA	NA	20 for 10% water 18 for 15% water	NA	THC reduced about 52% for 10% water 33% for 15% water
[50]	4 cylinder, air cooled, 1700 rpm and 2100 rpm	10 and 15%	NA	NA	NA	NA	NA	Reduction of 20% and 18% for 10% water and 15% water	NA	THC reduced 52% and 33% for 10% water and 15% water
[51]	6 cylinder, Caterpillar 3176 turbocharged engine, steady state operation	20% by mass	Purinox (commercial DE fuel)	NA	0.7% reduced	NA	NA	19% reduced	16% reduced	HC and CO emissions increased by 28% and 42%
[52]	Renault VI 620–45 (Euro 1) engine testing	13% by weight	NA	2–3%	Reduced 1–4%	NA	NA	Appr. 30% with reduction of 80% black smoke	Up to 50%	12% reduced HC
[53]	2.5L, 4cylinder.D.I. Ford engine, different load with 2500 rpm	20% by vol.	NA	NA	NA	NA	NA	Decrease Up to 60% with increased smoke	NA	HC and CO increased relatively low level

Engine Torque

Abu-Zaid, on his study of the effect of water content on the engine performance, has reported that the engine torque increases with an increase in the percentage of water in the emulsion [47]. According to Alahmer et al. [33], maximum torque was reported when the engine is fueled with a 5% water content by volume emulsified fuel. A reduction in torque with an emulsified fuel compared to the pure diesel fuel is reported by Nadeem et al. A relatively comparative torque is registered with 5% water by volume in the emulsion and gemini surfactant as an emulsifier used. The cause for the torque reduction in the emulsion is due to the reduction in heating value with an addition of water [32].

Engine Power

Abu-Zaid, on his study of the effect of water content on the engine performance, has reported that the engine power increases with an increase in the percentage of water in the emulsion [47] while Alahmer et al. reported that maximum power was achieved when the engine was fueled with a 5% water content by volume emulsified fuel [33]. On the other hand, Nadeem et al. have reported infinitesimal difference with the power output of the engine in the speed range of less than 4000 rpm. Even at 4000 rpm, pure diesel exhibited better power output compared to all emulsified fuels, with a relatively nearer performance with emulsified fuels using gemini surfactant [32]. A power loss of 7-8% was also reported by Barnes et al. on their application of WiDE with 10% water content by volume [71]. Since these results are based on different engine setups and methodologies, it is very difficult to explain the conflicting results reported on the engine power.

Engine Brake Specific Fuel Consumption

Brake specific fuel consumption (BSFC) was studied by Abu-Zaid by considering two cases. The first analysis has considered the total fuel as a sum of both the quantity of diesel and water resulting to an increased BSFC with an increase in the percentage of water in the emulsion. The second considered diesel alone as a total fuel and the analysis resulted with a decrease in BSFC with an increase in the percentage of water in the emulsion, the minimum value is reported to be at 20% water in the emulsion. The main reason for the reduction in BSFC is due to the secondary atomization of spray because of microexplosion [47]. This result was also shared by another publication by Kannan and Udayakumar on their experimental study on the effect of water percentage of water emulsified diesel fuel on BSFC. They have found that the BSFC of the engine decreases with an increase in the volume percentage of water in the emulsion, minimum value reported when the volume percentage of water was at 20%. This attribute, based on the report, is due to the displacement of diesel by water, resulting in less amount of diesel contained in the emulsion [44]. On the other hand, on separate study by Ghojel et al., 22–26% increase of BSFC was reported with emulsified diesel fuel of 13% water content by volume compared to diesel fuel [8]. Alahmer et al. have classified the effect of water percentage on the BSFC at high speed and low speed. According to their report, there has been an increase in BSFC with an increase in the percentage of water in the emulsion when the engine was at higher speed. There was no significant effect reported on BSFC with an increase in the percentage of water in the emulsion when the engine was at lower speed. Lowest engine BSFC was also reported with pure diesel fuel compared to emulsified fuels, with 15% water content emulsion taking the highest value [33]. The main factor attributing to this situation according to the authors is due to the displacement of diesel fuel with the amount of water added, which further will facilitate the fuel burning in the precombustion. An increase of BSFC in the range of 2–7% was reported by Barnes et al. on their study of effect of water blended fuel on the performance and emissions of a city bus engine considering a 10% water content by volume [71]. Armas et al. has also reported an increase in brake specific fuel consumption with 10% water content by volume in the emulsified fuel compared to pure diesel fuel [7].

While Abu-Zaid [47] and Kannan and Udayakumar [44] both reported improvement in BSFC with an increase in the percentage of water content in the emulsion, negative effect on BSFC was also reported [8, 32, 33, 71]. Higher BFSC irrespective of engine loading was reported by Alam Fahd et al. [11].

Engine Brake Thermal Efficiency

Kannan and Udayakumar have experimentally studied the effect of water percentage of water emulsified diesel fuel on brake thermal efficiency. They have found that the brake thermal efficiency of the engine increases with an increase in the volume percentage of water in the emulsion. This attribute as reported by the authors is due to an increase in expansion work and reduction in compression works as a result of expansion of water vapors [44]. Slight improvement in thermal efficiency was also reported by Armas et al. and Ghojel et al., 3.5% increase in brake thermal efficiency was reported for the study of engine fueled with 20% water in the emulsion according to Abu-Zaid [47]. Alahmer et al. reported that maximum brake thermal efficiency was achieved when the engine is fueled with a 5% water content by volume emulsified fuel [33]. with a static injection timing of 23° BTDC; Subramanian [46] compared the effects of WiDE and direct injection of water into the manifold and found WiDE to be more effective than injection of water with regard to brake thermal efficiency.

EFFECT OF WIDE ON EMISSIONS

The introduction of water by the emulsification process has many effects on the combustion process that have direct consequences on the pollutant formation. Vaporization of water due to heat absorption from its surroundings will lower the local high temperature resulting in the reduction of NO_x [5–8, 21, 26, 33, 44]. Alahmer et al. on their study of water emulsion on the performance and emission have reported that at low amount of water addition, the amount of emitted NO and NO_x increases, but at high water content, the amount of emitted NO and NO_x decreases [33]. Furthermore, Kannan and Udayakumar have mathematically modeled nitric oxide formation in single cylinder direct injection diesel engine using diesel-water Emulsion [72]. Based on their results, it was found that 18% and 21.5% of reduction in NO was achieved with 10% and 20% dilution of diesel with water, respectively. On their experimental investigation in another literature, the same authors have reported that 10% and 25% reduction of NO_x in a single cylinder diesel engine for 10% and 20% water in the emulsion was observed, respectively [44]. Ghojel and coworkers have reported

29–37% reduction of NO_x emissions when operating on diesel oil emulsion of 13% water content by volume [8]. Another experimental and numerical study conducted by Samec et al. reported a reduction of 20% and 18% NO_x emission compared to pure diesel fuel with 10% and 15% water content in the emulsion, respectively [49, 50]. A decrease of NO_x emission of 9% was reported by Barnes et al. on their study of effect of water blended fuel on the performance and emissions of a city bus engine considering a 10% water content by volume [71].

Likewise, there is also a microexplosion phenomenon as it has been discussed in detail. The effect of microexplosion is to facilitate the mixing process, in turn, it will reduce reaction time. Furthermore, the reduction in maximum local temperature also reduces the reaction rate. These combined effects reduce the formation of particulate matter and soot [7, 21] and total hydrocarbon [7, 8, 44] in the exhaust. The HC is also further reduced with the effect of OH radical that is dissociated from water [7, 44]. Ghojel et al. on their study of performance, emission, and heat release characteristics of direct injection diesel engine using diesel oil emulsion, have reported 60–90% reductions of HC emissions when operating on diesel oil emulsion of 13% water contents by volume compared to the base fuel [8]. Samec et al. reported a reduction of 52% and 33% total hydrocarbon emissions; and 68% and 75% reduction of soot emission compared to pure diesel fuel with 10% and 15% water content in the emulsion, respectively [49, 50]. Barnes et al. have reported a 20% decrease in PM emission on their study of effect of water blended fuel on the performance and emission of a city bus engine considering a 10% water content by volume [71]. Reduction in exhaust temperature and less CO has been reported for all engine loading conditions [11] but higher CO at low load, low speed was significantly reduced at higher engine rpm. At low load conditions of HSDI engine for 25.6% water to fuel ratio, NO_x is most often reduced upto 50% with 94% reduction in PM [45]. On the whole, WiDE is more effective in reducing NO and smoke level at low engine loads [46]. On the other hand, it is reported that there is an increase of CO_2 [33] and CO [5] emissions with water-in-diesel emulsion compared to the base diesel fuel. This is because of excess oxygen in the combustion mixture.

Armas et al. investigated the effect of 10% water addition with diesel on the emission levels of NO_x, total hydrocarbons (THC), soot, particulate matter (PM), and its composition [7]. There is a relative reduction in most of the pollutant emissions when the engine is operated with 10% water-in-diesel emulsion agreed with most of the literature in this area. According to a report by Sadler [73], an application of 13% water content (not mentioned whether by volume or mass percentage) in the emulsified fuel in UK has brought 13% and 25% reduction of NO_x and PM, respectively.

According to a report by Nadeem et al., [32] on their study performance and emission using conventional and gemini surfactant stabilized emulsified fuels, lowest PM, NO_x, and CO were produced by the engine when it was operated using emulsified fuel containing 15% water contents with gemini surfactants. Lin and Wang [57] on their study of engine performance and emissions characteristics using a three-phase emulsion prepared by two-stage emulsification method have reported an increase in CO_2 and CO emissions and decrease in O_2 and NO_x emissions with the emulsified fuel compared to neat diesel fuel. On their comparison between two-phase and three phase emulsions, three phase emulsion fuel has registered lower CO and NO_x emissions. A similar experimental study has been conducted by Lin and Chen [30] on a four cylinder diesel engine to compare fuel property and emission characteristics of two-phase and three-phase emulsions prepared by ultrasonic vibrator and mechanical homogenizer. They have reported results for NO, CO, CO_2, O_2, and smoke opacity. Largest content of NO was emitted when the engine was fueled with neat diesel fuel while three phase emulsion fuel prepared by mechanical homogenizer had lowest NO emission. With regard to CO emission, lower emission was registered with a two-phase emulsion fuel prepared by ultrasonic vibrator. A similar trend has been observed by CO_2 and O_2 emissions with all fuel types. Highest smoke opacity was registered with neat diesel fuel while lowest emission was observed with three phase emulsion fuel prepared by mechanical homogenizer.

CONCLUSIONS AND FUTURE RECOMMENDATION

WiDE fuel has become the best alternative fuel to substitute diesel fuel in both transport and stationary CI engines. The driving force for the growing interest to this type of fuel is simultaneous reduction of both

NO_x and particulate matters. This occurs as a result of the reduction in peak cylinder temperature and secondary atomization by a further breakup of fuel spray due to microexplosion. Although score of research have been conducted both experimentally and numerically outside the engine, studies of its effects inside the combustion chamber were quite few. Experimental investigation about the effect of various surfactants in the WiDE on engine performance and pollutant formation is not known. This review paper emphasise the research gap to investigate the effects of various surfactants with several blends of emulsified fuel on the combustion characteristics, emission formation processes, and engine behaviours also to determine the pollution formation suppression capability of the emulsified fuels by in-depth combustion characteristics analysis.

It is also equally important to select the suitable emulsification technique, optimised speed and agitation time in order to achieve stable emulsion.

There have been inconsistent results reported by different researchers with regard to the effect of water content on the engine combustion characteristics. Besides, all the reports are based on different engine setups and methodologies. As a result, an optimum percentage of water content in the emulsion cannot be drawn. But it can be concluded that water content ranging from 5–40% by volume in the emulsion has been utilized in the experimental and numerical investigation.

There was a common agreement by most of the researchers on the report of the effect of water content on the simultaneous reduction of both NO_x and particulate matter. The inconsistency was on the percentage amount reduction compared to pure diesel. Up to 37% reduction NO_x and 90% reduction in particulate matter were reported by different researchers. A systematic approach of studying the optimization of water content in the emulsion for best engine performance and emission by both experimental and numerical investigations is necessary so that it can give the best recommendations for the commercialization of WiDE as an alternative source of energy for the future diesel engines.

REFERENCES

1.   C. Lin and K. Wang, "The fuel properties of three-phase emulsions as an alternative fuel for diesel engines," Fuel, vol. 82, no. 11, pp.

1367–1375, 2003.

2. L. Xing-Cai, Y. Jian-Guang, Z. Wu-Gao, and H. Zhen, "Effect of cetane number improver on heat release rate and emissions of high speed diesel engine fueled with ethanol-diesel blend fuel," Fuel, vol. 83, no. 14-15, pp. 2013–2020, 2004.

3. A. Farfaletti, C. Astorga, G. Martini, et al., "Effect of water/fuel emulsions and a cerium-based combustion improver additive on HD and LD diesel exhaust emissions," Environmental Science & Technology, vol. 39, no. 17, pp. 6792–6799, 2005.

4. Directives and regulations-motor vehicles,http://ec.europa. eu/enterprise/sectors/automotive/documents/directives/motor-vehicles/.

5. A. Lif and K. Holmberg, "Water-in-diesel emulsions and related systems," Advances in Colloid and Interface Science, vol. 123-126, pp. 231–239, 2006.

6. T. Kadota and H. Yamasaki, "Recent advances in the combustion of water fuel emulsion," Progress in Energy and Combustion Science, vol. 28, no. 5, pp. 385–404, 2002.

7. O. Armas, R. Ballesteros, F. J. Martos, and J. R. Agudelo, "Characterization of light duty diesel engine pollutant emissions using water-emulsified fuel," Fuel, vol. 84, no. 7-8, pp. 1011–1018, 2005.

8. J. Ghojel, D. Honnery, and K. Al-Khaleefi, "Performance, emissions and heat release characteristics of direct injection diesel engine operating on diesel oil emulsion," Applied Thermal Engineering, vol. 26, no. 17-18, pp. 2132–2141, 2006.

9. X. T. Tran and J. Ghojel, "Impact of introducing water into the combustion chamber of diesel engines on emissions-an overview," in Proceedings of the 5th Asia pacific Conference on Combustion, The University of Adelaide, Adelaide Australia, July 2005.

10. Transportation and Air Quality, Impacts of Lubrizol's PuriNOx Water/Diesel Emulsion on Exhaust Emissions from Heavy-Duty Engines, Assessment and Standards Division Office of Transportation and Air Quality U.S. Environmental Protection Agency, Ann Arbor, Mich, USA, 2002.

11. E. Alam Fahd, Y. Wenming, P. Lee, S. Chou, and C. Yap, "Experimental investigation of the performance and emission characteristics of direct injection diesel engine by water emulsion diesel under varying engine load condition," Applied Energy, vol. 102, pp. 1042–1049, 2013.

12. H.-Z. Sheng, Z.-P. Zhang, and C.-K. Wu, "Study of atomization and micro-explosion of water in diesel fuel emulsion droplets in spray with in a high temperature, high pressure bomb," in Proceedings of the International Symposium on Diagnostics and Modeling of Combustion in Internal Combustion Engines, Kyoto, Japan, 1990.

13. M. Abu-Zaid, "An experimental study of the evaporation characteristics of emulsified liquid droplets," Heat and Mass Transfer/Waerme- und Stoffuebertragung, vol. 40, no. 9, pp. 737–741, 2004.

14. H. Tanaka, H. Yamasaki, S. Teraji, D. Segawa, and T. Kadota, "Effects of fuel properties, water contents and surface temperatures on micro-explosion of emulsion droplets burning on a hot surface," Transactions of the Japan Society of Mechanical Engineers B, vol. 71, no. 702, pp. 690–695, 2005.

15. H. Tanaka, T. Kadota, D. Segawa, S. Nakaya, and H. Yamasaki, "Effect of ambient pressure on micro-explosion of an emulsion droplet evaporating on a hot surface," JSME International Journal B, vol. 49, no. 4, pp. 1345–1350, 2007.

16. M. Tsue, T. Kadota, D. Segawa, and H. Yamasaki, "Statistical analysis of onset of microexplosion for an emulsion droplet," Symposium (International) on Combustion, vol. 26, no. 1, pp. 1629–1635, 1996. ·

17. H. Watanabe, Y. Suzuki, T. Harada, Y. Matsushita, H. Aoki, and T. Miura, "An experimental investigation of the breakup characteristics of secondary atomization of emulsified fuel droplet," Energy, vol. 35, no. 2, pp. 806–813, 2010.

18. I. Jeong, K. Lee, and J. Kim, "Characteristics of auto-ignition and micro-explosion behavior of a single droplet of water-in-fuel," Journal of Mechanical Science and Technology, vol. 22, no. 1, pp. 148–156, 2008.

19. I. C. Jeong and K. H. Lee, "Auto-ignition and micro-explosion behaviors of droplet arrays of water-in-fuel emulsion,"

International Journal of Automotive Technology, vol. 9, no. 6, pp. 735–740, 2008. ·

20. Y. Morozumi and Y. Saito, "Effect of physical properties on microexplosion occurrence in water-in-oil emulsion droplets," Energy and Fuels, vol. 24, no. 3, pp. 1854–1859, 2010

21. T. Yatsufusa, T. Kumura, Y. Nakagawa, and Y. Kidoguch, "Advantage of using water-emulsified fuel on combustion and emission characteristics," in Proceedings of the European Combustion Meeting, Vienna, Austria, 2009.

22. R. Ochoterena, A. Lif, M. Nydén, S. Andersson, and I. Denbratt, "Optical studies of spray development and combustion of water-in-diesel emulsion and microemulsion fuels," Fuel, vol. 89, no. 1, pp. 122–132, 2010.

23. J. Shao and Y. Yan, "Digital imaging based measurement of diesel spray characteristics," IEEE Transactions on Instrumentation and Measurement, vol. 57, no. 9, pp. 2067–2073, 2008.

24. Y.-T. Han, K.-H. Lee, and K.-D. Min, "A study on the measurement of temperature and soot in a constant-volume chamber and a visualized diesel engine using the two-color method," Journal of Mechanical Science and Technology, vol. 23, no. 11, pp. 3114–3123, 2010.

25. R. S. G. Baert, P. J. M. Frijters, B. Somers, C. Luijten, and W. d. Boer, "Design and operation of a high pressure, high temperature cell for HD diesel spray diagnostics: guidelines and results," SAE Transactions2009-01-0649, 2009.

26. J. Park, K. Huh, and K. Park, "Experimental study on the combustion characteristics of emulsified diesel in a rcem," in Proceedings of the Seoul FISITA World Automotive Congress, Seoul, Republic of Korea, June 2000.

27. J. I. Ghojel and X. Tran, "Ignition characteristics of diesel-water emulsion sprays in a constant-volume vessel: effect of injection pressure and water content," Energy and Fuels, vol. 24, no. 7, pp. 3860–3866, 2010.

28. M. P. Ashok and C. G. Saravanan, "Combustion characteristics of compression engine driven by emulsified fuel under various fuel injection angles," Journal of Energy Resources Technology, vol. 129, no. 4, pp. 325–331, 2007.

29. M. P. Ashok and C. G. Saravanan, "The performance and emission characteristics of emulsified fuel in a direct injection diesel engine," Proceedings of the Institution of Mechanical Engineers D, vol. 221, no. 7, pp. 893–900, 2007.

30. C.-Y. Lin and L.-W. Chen, "Comparison of fuel properties and emission characteristics of two- and three-phase emulsions prepared by ultrasonically vibrating and mechanically homogenizing emulsification methods," Fuel, vol. 87, no. 10-11, pp. 2154–2161, 2008.

31. C.-Y. Lin and K.-H. Wang, "Diesel engine performance and emission characteristics using three-phase emulsions as fuel," Fuel, vol. 83, no. 4-5, pp. 537–545, 2004.

32. M. Nadeem, C. Rangkuti, K. Anuar, M. R. U. Haq, I. B. Tan, and S. S. Shah, "Diesel engine performance and emission evaluation using emulsified fuels stabilized by conventional and gemini surfactants," Fuel, vol. 85, no. 14-15, pp. 2111–2119, 2006.

33. A. Alahmer, J. Yamin, A. Sakhrieh, and M. A. Hamdan, "Engine performance using emulsified diesel fuel," Energy Conversion and Management, vol. 51, no. 8, pp. 1708–1713, 2010.

34. B. Andrea, L. Renxian, and B. Konstantinos, "Influence of water-diesel fuel emulsions and EGR on combustion and exhaust emissions of heavy duty DI-diesel engines equipped with common-rail injection system," SAE Transactions, vol. 112, pp. 2244–2260, 2003.

35. C. A. Canfield, Effects of diesel-water emulsion combustion diesel engine NOx emissions [M.S. thesis], Mechanical Engineering, University of Florida, Gainesville, Fla, USA, 1999.

36. Y. Zeng and C.-F. Lee, "Modeling droplet breakup processes under micro-explosion conditions," inProceedings of the 31st International Symposium on Combustion, vol. 31, pp. 2185–2193, August 2006.

37. H. Watanabe, Y. Matsushita, H. Aoki, and T. Miura, "Numerical simulation of emulsified fuel spray combustion with puffing and micro-explosion," Combustion and Flame, vol. 157, no. 5, pp. 839–852, 2010.

38. D. Tarlet, J. Bellettre, M. Tazerout, and C. Rahmouni, "Prediction of micro-explosion delay of emulsified fuel droplets," International Journal of Thermal Sciences, vol. 48, no. 2, pp. 449–460, 2009.

39. P. C. L. Clercq, B. Noll, and M. Aigner, Modeling Evaporation and Secondary Atomization of Water-in-Multicomponent Oil Emulsion Droplets, DLR, German Aerospace Center, Institute of Combustion Technology, Stuttgart, Germany, 2005.

40. W. B. Fu, L. Y. Hou, L. Wang, and F. H. Ma, "A unified model for the micro-explosion of emulsified droplets of oil and water," Fuel Processing Technology, vol. 79, no. 2, pp. 107–119, 2002. ·

41. J. Ghojel and D. Honnery, "Heat release model for the combustion of diesel oil emulsions in di diesel engines," Applied Thermal Engineering, vol. 25, no. 14-15, pp. 2072–2085, 2005. · ·

42. P. Eckert, A. Velji, and U. Spicher, "Numerical investigations of fuel-water emulsion combustion in DI-diesel Eng," in Proceedings of the International Council on Combution Engines (CIMAC ‹07), Vienna, Austria, 2007.

43. J. Weber, N. Peters, H. Bockhorn, and R. Pittermann, "Numerical simulation of the evolution of the soot particle size distribution in a DI diesel engine using an emulsified fuel of diesel-water," SAE Transactions, vol. 113, pp. 2217–2233, 2004.

44. K. Kannan and M. Udayakumar, "NOx and HC emission control using water emulsified diesel in single cylinder diesel engine," Journal of Engineering and Applied Sciences, vol. 4, no. 8, pp. 59–62, 2009.

45. A. Maiboom and X. Tauzia, "NOx and PM emissions reduction on an automotive HSDI Diesel engine with water-in-diesel emulsion and EGR: an experimental study," Fuel, vol. 90, no. 11, pp. 3179–3192, 2011. · ·

46. K. A. Subramanian, "A comparison of water-diesel emulsion and timed injection of water into the intake manifold of a diesel engine for simultaneous control of NO and smoke emissions," Energy Conversion and Management, vol. 52, no. 2, pp. 849–857, 2011. · ·

47. M. Abu-Zaid, "Performance of single cylinder, direct injection Diesel engine using water fuel emulsions," Energy Conversion and Management, vol. 45, no. 5, pp. 697–705, 2004. · ·

48. K. Park, I. Kwak, and S. Oh, "The effect of water emulsified fuel on a motorway-bus diesel engine," KSME International Journal, vol. 18, no. 11, pp. 2049–2057, 2004.

49. N. Samec, B. Kegl, and R. W. Dibble, "Numerical and experimental study of water/oil emulsified fuel combustion in a diesel engine," Fuel, vol. 81, no. 16, pp. 2035–2044, 2002. · ·

50. N. Samec, Z. Dobovisek, and A. Hribernik, "The Effect of Water emulsified in diesel fuel on diesel fuel on exhaust emission," Goriva i Maziva, vol. 39, pp. 377–392, 2000.

51. A. C. Matheaus, T. W. Ryan, D. Daly, D. A. Langer, and M. P. B. Musculus, "Effects of PuriNOx water-diesel fuel emulsions on emissions and fuel economy in a heavy-duty diesel engine," SAE Transactions 2002-01-2891, 2002.

52. F. Barnaud, P. Schmelzle, and P. Schulz, "AQUAZOLE: an original emulsified water-diesel fuel for heavy-duty applications," SAE Transactions 2000-01-1861, 2000.

53. M. A. A. Nazha, H. Rajakaruna, and S. A. Wagstaff, "The use of emulsion, water induction and EGR for controlling diesel engine emissions," SAE Transactions 2001-01-1941, 2001.

54. M. Fingas and B. Fieldhouse, "Formation of water-in-oil emulsions and application to oil spill modelling," Journal of Hazardous Materials, vol. 107, no. 1-2, pp. 37–50, 2004. · ·

55. J. Jiao and D. J. Burgess, "Rheology and stability of water-in-oil-in-water multiple emulsions containing span 83 and tween 80," AAPS PharmSci, vol. 5, no. 1, p. e7, 2003. · ·

56. Laboratory 6-Characteristics of Surfactants and Emulsions, http://pharmacy.wilkes.edu/kibbeweb/lab7.html.

57. C.-Y. Lin and K.-H. Wang, "Diesel engine performance and emission characteristics using three-phase emulsions as fuel," Fuel, vol. 83, no. 4-5, pp. 537–545, 2004. · ·

58. C.-Y. Lin and L.-W. Chen, "Emulsification characteristics of three- and two-phase emulsions prepared by the ultrasonic emulsification method," Fuel Processing Technology, vol. 87, no. 4, pp. 309–317, 2006. · ·

59. Y. Hiroshi, T. Mitsuhiro, and K. Toshikazu, "Evaporation and combustion of emulsified fuel: onset of microexplosion," Japan Society of Mechanical Engineers International Journal, vol. 36, pp. 677–681, 1993.

60. M. Tsue, H. Yamasaki, T. Kadota, D. Segawa, and M. Kono, "Effect of gravity on onset of microexplosion for an oil-in-water emulsion

droplet," Symposium (International) on Combustion, vol. 2, pp. 2587–2593, 1998.

61. M. Y. E. Selim and M. T. Ghannam, "Combustion study of stabilized water-in-diesel fuel emulsion,"Energy Sources A, vol. 32, no. 3, pp. 256–274, 2010. · ·

62. G. Chen and D. Tao, "An experimental study of stability of oil-water emulsion," Fuel Processing Technology, vol. 86, no. 5, pp. 499–508, 2005. · ·

63. H. Watanabe, T. Harada, Y. Matsushita, H. Aoki, and T. Miura, "The characteristics of puffing of the carbonated emulsified fuel," International Journal of Heat and Mass Transfer, vol. 52, no. 15-16, pp. 3676–3684, 2009. · ·

64. M. Ghannam and M. Y. Selim, "Stability behavior of water-in-diesel fuel emulsion," Petroleum Science and Technology, vol. 27, no. 4, pp. 396–411, 2009.

65. V. Ganesan, Internal Combustion Engines, Tata McGraw-Hill, New Delhi, India, 1994.

66. S. C. Siegmund, M. L. Storbeck, J. B. Cross, and H. S. Fogler, "Physical properties of water in fuel oil emulsions (density and bulk compressibility)," Journal of Chemical and Engineering Data, vol. 25, no. 1, pp. 72–74, 1980.

67. F. Y. Hagos, A. R. A. Aziz, and I. M. Tan, "Water-in-diesel emulsion and its micro-explosion phenomenon-review," in Proceedings of the IEEE 3rd International Conference on Communication Software and Networks (ICCSN ‹11), pp. 314–318, Bali, Indonesia, May 2011. · ·

68. W. Fu, J. Gong, and L. Hou, "There is no micro-explosion in the diesel engines fueled with emulsified fuel," Chinese Science Bulletin, vol. 51, no. 10, pp. 1261–1265, 2006. · ·

69. N. J. Marrone, I. M. Kennedy, and F. L. Dryer, "Internal phase size effects on combustion of emulsions,"Combustion Science and Technology, vol. 33, no. 5-6, pp. 299–307, 1983.

70. M. Y. E. Selim and S. M. S. Elfeky, "Effects of diesel/water emulsion on heat flow and thermal loading in a precombustion chamber diesel engine," Applied Thermal Engineering, vol. 21, no. 15, pp. 1565–1582, 2001. · ·

71. A. Barnes, D. Duncan, J. Marshall, A. Psaila, J. Chadderton, and A. Eastlake, "Evaluation of water-blend fuels in a city bus and an assessment of performance with emission control devices," in Proceedings of the Better air Quality Motor Vehicle Control & Technology Workshop 2000, 2000.

72. K. Kannan and M. Udayakumar, "Modeling of nitric oxide formation in single cylinder direct injection diesel engine using diesel-water emulsion," American Journal of Applied Sciences, vol. 6, no. 7, pp. 1313–1320, 2009. · ·

73. L. Sadler, The Air Quality Impact of Water-Diesel Emulsion Fuel (WDE) and Selective Catalytic Reduction (SCR) Technologies, Mayor of London, London, UK, 2003.

Chapter 4

Current Progress on Bio-based Polymers and their Future Trends

Ramesh P Babu[1, 2], Kevin O'Connor[3], and
Ramakrishna Seeram[4, 5, 6]

[1]Centre for Research Adoptive Nanostructures and Nano Devices, Trinity College, Dublin 2, Ireland

[2]School of Physics, Trinity College Dublin, Dublin 2, Ireland

[3]School of Biomolecular and Biomedical Sciences, Centre for Synthesis and Chemical Biology, UCD Conway Institute, and Earth Institute, University College Dublin, Belfield, Dublin 4, Ireland

[4]NUSNNI, National University of Singapore, 2 Engineering Drive 3, Singapore, 117581, Singapore

[5]Institute of Materials Research and Engineering, Singapore, 117602, Singapore

[6]Jinan University, Guangzhou, China

ABSTRACT

This article reviews the recent trends, developments, and future applications of bio-based polymers produced from renewable resources. Bio-based polymers are attracting increased attention due to environmental concerns and the realization that global petroleum resources are finite. Bio-based polymers not only replace existing polymers in a number of applications but also provide new combinations of properties for new applications. A range of bio-based polymers are presented in this review, focusing on general methods of production, properties, and commercial applications. The review examines the technological and future challenges discussed in bringing these materials to a wide range of applications, together with potential solutions, as well as discusses the major industry players who are bringing these materials to the market.

REVIEW

Introduction

Bio-based polymers are materials which are produced from renewable resources. The terms bio-based polymers and biodegradable polymers are used extensively in the literature, but there is a key difference between the two types of polymers. Biodegradable polymers are defined as materials whose physical and chemical properties undergo deterioration and completely degrade when exposed to microorganisms, carbon dioxide (aerobic) processes, methane (anaerobic processes), and water (aerobic and anaerobic processes). Bio-based polymers can be biodegradable (e.g., polylactic acid) or nondegradable (e.g., biopolyhethylene). Similarly, while many bio-based polymers are biodegradable (e.g., starch and polyhydroxyalkanoates), not all biodegradable polymers are bio-based (e.g., polycaprolactone).

Bio-based polymers still hold a tiny fraction of the total global plastic market. Currently, biopolymers share less than 1% of the total market. At the current growth rate, it is expected that biopolymers will account for just over 1% of polymers by 2015 (Doug 2010).

The worldwide interest in bio-based polymers has accelerated in recent years due to the desire and need to find non-fossil fuel-based polymers. As indicated by ISI Web of Sciences and Thomas Innovations, there is a tremendous increase in the number of publication citations on bio-based polymers and applications in recent years, as shown in Figure 1 (Chen and Martin 2012).

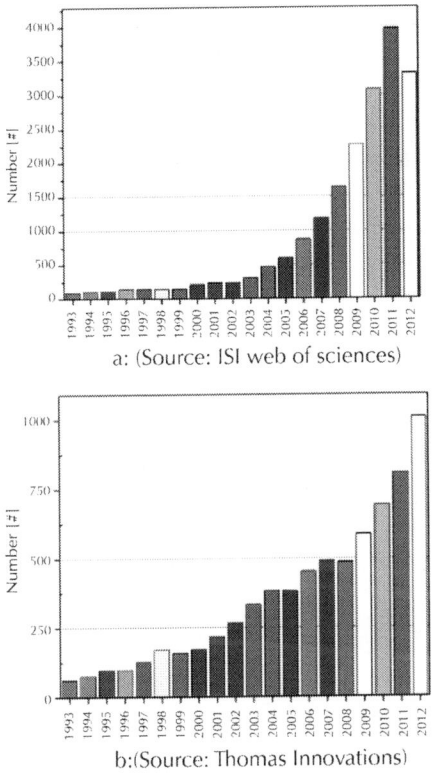

a: (Source: ISI web of sciences)

b:(Source: Thomas Innovations)

Figure 1: Citation trends of (a) publications and (b) patents on bio-based polymers in recent years.

Bio-based polymers offer important contributions by reducing the dependence on fossil fuels and through the related positive environmental impacts such as reduced carbon dioxide emissions. The legislative landscape is also changing where bio-based products are being favored through initiatives such as the *Lead Market Initiative* (European Union) and *BioPreferred* (USA). As a result, there is a worldwide demand for replacing petroleum-derived raw materials with renewable resource-based raw materials for the production of polymers.

The first generation of bio-based polymers focused on deriving polymers from agricultural feedstocks such as corn, potatoes, and other carbohydrate feedstocks. However, the focus has shifted in recent years due to a desire to move away from food-based resources and significant breakthroughs in biotechnology. Bio-based polymers similar to conventional polymers are produced by bacterial fermentation processes by synthesizing the building blocks (monomers) from renewable resources, including lignocellulosic biomass (starch and cellulose), fatty acids, and organic waste. Natural bio-based polymers are the other class of bio-based polymers which are found naturally, such as proteins, nucleic acids, and polysaccharides (collagen, chitosan, etc.). These bio-based polymers have shown enormous growth in recent years in terms of technological developments and their commercial applications. There are three principal ways to produce bio-based polymers using renewable resources:

1. Using natural bio-based polymers with partial modification to meet the requirements (e.g., starch)

2. Producing bio-based monomers by fermentation/conventional chemistry followed by polymerization (e.g., polylactic acid, polybutylene succinate, and polyethylene)

3. Producing bio-based polymers directly by bacteria (e.g., polyhydroxyalkanoates).

In this paper, an overview of bio-based polymers made from renewable resources and natural polymers derived from plant and animal origins is presented. The review will focus on the preparation, properties, applications, and future trends for bio-based polymers. This paper discusses the use of renewable resources such as lignocellulosic biomass to create monomers and polymers that can replace petroleum-based polymers, such as polyester, polylactic acids, and other natural bio-based polymers, which are presented in Figure 2.

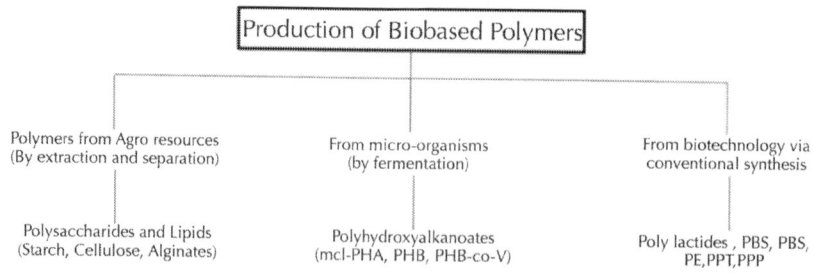

Figure 2: Most common categories of bio-based polymers produced by various processes. From Luc and Eric (2012).

Polylactic Acid

Polylactic acid (PLA) has been known since 1845 but not commercialized until early 1990. PLA belongs to the family of aliphatic polyesters with the basic constitutional unit lactic acid. The monomer lactic acid is the hydroxyl carboxylic acid which can be obtained via bacterial fermentation from corn (starch) or sugars obtained from renewable resources. Although other renewable resources can be used, corn has the advantage of providing a high-quality feedstock for fermentation which results in a high-purity lactic acid, which is required for an efficient synthetic process. l-lactic acid or d-lactic acid is obtained depending on the microbial strain used during the fermentation process.

PLA can be synthesized from lactic acid by direct polycondensation reaction or ring-opening polymerization of lactide monomer. However, it is difficult to obtain high molecular weight PLA via polycondensation reaction because of water formation during the reaction. Nature Works LLC (previously Cargill Dow LLC) has developed a low-cost continuous process for the production of PLA (Erwin et al. 2007). In this process, low molecular weight pre-polymer lactide dimers are formed during a condensation process. In the second step, the pre-polymers are converted into high molecular weight PLA via ring-opening polymerization with selected catalysts. Depending on the ratio and stereochemical nature of the monomer (l or d), various types of PLA and PLA copolymers can be obtained. The final properties of PLA produced are highly dependent on the ratio of the d and l forms of the lactic acid which are listed in Table 1 for various blend ratios (Garlotta 2001).

Table 1: Variation in glass transition and melting temperature of PLA with various ratios of L-monomer composition

Copolymer ratio	Glass transition (T_g), C	Melting temperature (T_m), C
100:0 (l/dl)-PLA	63	178
95:5 (l/dl)-PLA	59	164
90:10 (l/dl)-PLA	56	150
85:15 (l/dl)-PLA	56	140
80:20 (l/dl)-PLA	56	125

Babu *et al.*

Babu *et al.* Progress in Biomaterials 2013 2:8, doi:10.1186/2194-0517-2-8

PLA is a commercially interesting polymer as it shares some similarities with hydrocarbon polymers such as polyethylene terephthalate (PET). It has many unique characteristics, including good transparency, glossy appearance, high rigidity, and ability to tolerate various types of processing conditions.

PLA is a thermoplastic polymer which has the potential to replace traditional polymers such as PET, PS, and PC for packaging to electronic and automotive applications (Majid et al. 2010). While PLA has similar mechanical properties to traditional polymers, the thermal properties are not attractive due to low Tg of 60°C. This problem can be overcome by changing the stereochemistry of the polymer and blending with other polymers and processing aids to improve the mechanical properties, e.g., varying the ratio of l and d isomer ratio strongly influences the crystallinity of the final polymer. However, much more work is required to improve the properties of PLA to suit various applications.

Currently, Nature Works LLC, USA, is the major supplier of PLA sold under the brand name Ingeo, with a production capacity of 100,000 ton/year. There are other manufactures of PLA based in the USA, Europe, China, and Japan developing various grades of PLA suitable for different industrial sectors such as automobile, electronics, medical devices, and commodity applications, which are mentioned in Table 2) (Doug 2010; Ravenstijn 2010).

Table 2: Global suppliers of PLA

Company	Location	Brand name	Production/planned capacity
			(kton/year)
Nature Works	USA	Ingeo	140 (by 2013)
Futerro	Belgium	Futerro	1.5 (by 2010)
Tate & Lyle	Netherlands	Hycail	0.2 (by 2012)
Purac	Netherlands	Purasorb	0.05
Hiusan Biosciences	China	Hisun	5
Jiangsu Jiulding	China		5
Teijin	Japan	Biofront	1
Toyobo	Japan	Vylocol	0.2
Synbra	Netherlands	Biofoam	50

Babu *et al.*
Babu *et al. Progress in Biomaterials* 2013 2:8, doi:10.1186/2194-0517-2-8

PLA is widely used in many day-to-day applications. It has been mainly used in food packing (including food trays, tableware such as plates and cutlery, water bottles, candy wraps, cups, etc.). Although PLA has one of the highest heat resistances and mechanical strengths of all bio-based polymers, it is still not suitable for use in electronic devices and other engineering applications. NEC Corporation (Japan) recently produced a PLA with carbon and kenaf fibers with improved thermal and flame retardancy properties. Fujitsu (Japan) developed a polycarbonate blend with PLA to make computer housings. In recent years, PLA has been employed as a membrane material for use in automotive and chemical industry.

The ease of melt processing has led to the production of PLA fibers, which are increasingly accepted in a wide variety of textiles from dresses to sportswear, furnishing to drapes, and soft nonwoven baby wipes to tough landscape textiles. These textiles can outperform traditional textiles made from synthetic counterparts. Bioresorbable scaffolds produced with PLA and various PLA blends are used in implants for growing living cells. The US Food and Drug Administration (FDA) has approved the use of PLA for certain human clinical applications (Dorozhkin 2009; Garlotta2001). In addition, PLA-based materials have been used for bone support splints. Applications of PLA-based polymers in various fields are listed in Table 3.

Table 3: Application of PLA and their blends in various fields

Polymer	Applications	Reference
PLGA/PGA	Ovine pulmonary valve replacement	Williams et al. 1999; Sodian et al. 1999, 2000; Cheng et al. 2009
PLA/chitosan PLA/PLGA/chitosan PLA	Drug carrier/drug release	Jeevitha and Kanchana 2013; Jayanth and Vinod 2012; Nagarwal et al. 2010; Chandy et al. 2000; Valantin et al.2003
PLGA and copolymers	Degradable sutures	Rajev 2000
PLA/HA composites	Porous scaffolds for cellular applications	Jung-Ju et al. 2012
PLA-CaP and PLGA-CaP	Bone fixation devices, plates, pins, screws, and wires, orthopedic applications	Huan et al. 2012

PDLLA	Coatings on metal implants	Schmidmaier et al. 2001
PLA/PLGA	Use in cell-based gene therapy for cardiovascular diseases, muscle tissues, bone and cartilage regeneration, and other treatments of cardiovascular and neurological conditions	Coutu et al. 2009; Kellomaki et al. 2000; Papenburg et al. 2009
PLA and PLA blends	Packaging films, commodity containers, electrical appliances, mobile phone housings, floor mats, automotive spare parts	Rafael et al. 2010
PLA	Textile applications	Gupta et al. 2007; Avinc and Akbar 2009

PLGA, polylactic acid-co-glycolic acid; CaP, calcium phosphates; HA, hydroxyapatite.

Babu *et al.*

Babu *et al. Progress in Biomaterials* 2013 2:8, doi:10.1186/2194-0517-2-8

Polyhydroxyalkanoates

Polyhydroxyalkanoates (PHAs) are a family of polyesters produced by bacterial fermentation with the potential to replace conventional hydrocarbon-based polymers. PHAs occur naturally in a variety of organisms, but microorganisms can be employed to tailor their production in cells. Polyhydroxybutyrate (PHB), the simplest PHA, was discovered in 1926 by Maurice Lemoigne as a constituent of the bacterium *Bacillus megaterium* (Lemoigne 1923).

PHA can be produced by varieties of bacteria using several renewable waste feedstocks. A generic process to produce PHA by bacterial fermentation involves fermentation, isolation, and purification from fermentation broth. A large fermentation vessel is filled with mineral medium and inoculated with a seed culture that contains bacteria. The feedstocks include cellulosics, vegetable oils, organic waste, municipal solid waste, and fatty acids depending on the specific PHA required. The carbon source is fed into the vessel until it is consumed and cell growth and PHA accumulation is complete. In general, a minimum of 48 h is required for fermentation time. To isolate and purify PHA, cells are concentrated, dried, and extracted with solvents such as acetone or chloroform. The residual cell debris is removed from the solvent containing dissolved PHA by solid-liquid separation process. The PHA is then precipitated by the addition of an alcohol (e.g., methanol) and recovered by a precipitation process (Kathiraser et al. 2007).

More than 150 PHA monomers have been identified as the constituents of PHAs (Steinbüchel and Valentin 1995). Such diversity allows the production of bio-based polymers with a wide range of properties, tailored for specific applications. Poly-3-hydroxybutyrate was the first bacterial PHA identified. It has received the greatest attention in terms of pathway characterization and industrial-scale production. It possesses similar thermal and mechanical properties to those of polystyrene and polypropylene (Savenkova et al. 2000). However, due to its slow crystallization, narrow processing temperature range, and tendency to 'creep', it is not attractive for many applications, requiring development in order to overcome these shortcomings (Reis et al. 2008). Several companies have developed PHA copolymers with typically 80% to 95% (R)-3-hydroxybutyric acid monomer and 5% to 20% of a second monomer in order to improve the properties of PHAs. Some specific examples of PHAs include the following:

- Poly(3HB): Poly(3-hydroxybutyrate)
- Poly(3HB-co-3HV):   Poly(3-hydroxybutyrate-co-3-hydroxyvalerate), PHBV
- Poly(3-HB-co-4HB):   Poly(3-hydroxybutyrate-co-4-hydroxybutyrate)
- Poly(3HB-co-3HH):   Poly(3-hydroxyoctanoate-co-hydroxyhexanoate)
- Poly(3HO-co-3HH):   Poly(3-hydroxyoctanoate-co-hydroxyhexanoate)
- Poly (4-HB): Poly (4-hydroxybutyrate).

The copolymer poly (3HB-co-3HV) has a much lower crystallinity, decreased stiffness and brittleness, and increased tensile strength and toughness compared to poly (3HB) while remaining biodegradable. It also has a higher melt viscosity, which is a desirable property for extrusion and blow molding (Hanggi 1995).

The first commercial plant for PHBV was built in the USA in a joint venture between Metabolix and Archer Daniels Midland. However, the joint venture between these two companies ended in 2012. Currently, Tianan Biologic Material Co. in China is the largest producer of PHB and PHB copolymers. Tianan's PHBV contains about 5% valerate which improves the flexibility of the polymer. Tainjin Green Biosciences, China, invested along with DSM to build a production plant with 10-kton/year capacity to produce PHAs for packing and biomedical applications (DSM press release 2008). The current global manufacturers of PHB-based polymers are listed in Table 4(Doug 2010; Ravenstijn 2010).

Table 4: Global suppliers of various types of PHAs

Company	Location	Brand name	Production/planned capacity (kton/year)
Bio-on	Italy	Minerv	10
Kaneka	Singapore		10 (by 2013)
Meredian	USA		13.5
Metabolix	USA	Mirel	50
Mitsubishi Gas Chemicals	Japan	Biogreen	0.05
PHB Industrial S/A	Brazil	Biocycle	0.05
Shenzen O'Bioer	China		
TEPHA	USA	ThephaFLEX/ ThephELAST	
Tianan Biological Materials	China	Enmat	2
Tianjin Green Biosciences	China	Green Bio	10
Tianjin Northern Food	China		
Yikeman Shandong	China		3

Babu *et al.*

Babu *et al. Progress in Biomaterials* 2013 2:8, doi:10.1186/2194-0517-2-8

PHA polymers are thermoplastic, and their thermal and mechanical properties depend on their composition. The Tg of the polymers varies from $-40°C$ to $5°C$, and the melting temperatures range from $50°C$ to $180°C$, depending on their chemical composition (McChalicher and Srienc2007). PHB is similar in its material properties to polypropylene, with a good resistance to moisture and aroma barrier properties. Polyhydroxybutyric acid synthesized from pure PHB is relatively brittle and stiff. PHB copolymers, which may include other fatty acids such as beta-hydroxyvaleric acid, may be elastic (McChalicher and Srienc 2007).

PHAs can be processed in existing polymer-processing equipment and can be converted into injection-molded components: film and sheet, fibers, laminates, and coated articles; nonwoven fabrics, synthetic paper products, disposable items, feminine hygiene products, adhesives, waxes, paints, binders, and foams. Metabolix has received FDA clearance for use of PHAs in food contact applications. These materials are suitable for a wide range of food packing applications including caps and closures, disposable items such as forks, spoons, knives, tubs, trays, and hot cup lids, and products such as housewares, cosmetics, and medical packaging (Philip et al. 2007).

PHA and its copolymers are widely used as biomedical implant materials. Various applications of PHA and their polymer blends are listed in Table 5. These include sutures, suture fasteners, meniscus repair devices, rivets, bone plates, surgical mesh, repair patches, cardiovascular patches, tissue repair patches, and stem cell growth. Changing the PHA composition allows the manufacturer to tune the properties such as biocompatibility and polymer degradation time within desirable time frames under specific conditions. PHAs can also be used in drug delivery due to their biocompatibility and controlled degradability. Only a few examples of PHAs have been evaluated for this type of applications, and it remains an important area for exploitation (Tang et al. 2008).

Table 5: Application of PHAs and their blends in various fields

PHA polymer type	Applications	Reference
P(3HB), P(3HB-co-3HHX) and blends	Scaffolds, nerve regeneration, soft tissue, artificial esophagus, drug delivery, skin regeneration, food additive	Yang et al. 2002; Chen and Qiong 2005; Bayram and Denbas 2008; Tang et al. 2008; Clarinval and Halleux2005
mcl-PHA/scl-PHA	Cardiac tissue engineering, drug delivery, cosmetics, drug molecules	Sodian et al. 2000; Wang et al. 2003; de Roo et al.2002; Zhao et al. 2003; Ruth et al. 2007
P(4HB) and P(3HO)	Heart valve scaffolds, food additive	Clarinval and Halleux 2005; Valappil et al. 2006
P(3HB-co-4HB), P(3HB-co-3HV)	Drug delivery, scaffolds, artificial heart values, patches to repair gastrointestinal tracts, sutures	Türesin et al. 2001; Williams et al. 1999; Chen et al.2008; Freier et al. 2002; Kunze et al. 2006; Volova et al. 2003
PHB, Mirel P103	Commodity applications, shampoo and cosmetic bottles, cups and food containers	Philip et al. 2007; Amass et al. 1998; Walle et al. 2001

Babu et al.

Babu et al. Progress in Biomaterials 2013 2:8, doi:10.1186/2194-0517-2-8

Polybutylene Succinate

Polybutylene succinate (PBS) is an aliphatic polyester with similar properties to those of PET. PBS is produced by condensation of succinic acid and 1,4-butanediol. PBS can be produced by either monomers derived from petroleum-based systems or the bacterial fermentation route. There are several processes for producing succinic acid from fossil fuels. Among them, electrochemical synthesis is a common process with high yield and low cost. However, the fermentation production of succinic acid has numerous advantages compared to the chemical process. Fermentation process uses renewable resources and consumes less energy compared to chemical process. Several companies (solely or in partnership) are now scaling bio-succinate production processes which have traditionally suffered from poor productivity and high downstream processing costs. Mitsubishi

Chemical (Japan) has developed biomass-derived succinic acid in collaboration with Ajinomoto to commercialize bio-based PBS. DSM and Roquette are developing a commercially feasible fermentation process for the production of succinic acid 1,4-butanediol and subsequent production of PBS. Myriant and Bioamber have developed a fermentation technology to produce monomers. There are several companies around the world developing technologies for the production of PBS, as listed in Table 6, including North America and China (Doug 2010; Ravenstijn2010).

Table 6: Global producers of PBS

Company	Location	Brand name/polymer type	Production/planned capacity (kton/year)
BASF	Germany	PBS	
Dupont de Nemours	USA	PBST	
Hexing Chemical	China	PBS	3
Ube	Japan	NA	NA
IPC-CAS	China	PBS, PBSA	5
IRE Chemical	Korea	Enpol, PBS, PBSA	3.5
Kingfa	China	PBSA	1
Mitsubishi Gas Chemical	Japan	PBS, PES, PBSLa	3
Showa	Japan	Bionelle PBS, PBSA, PBS	3
SK Chemicals	Korea	Skygreen	NA
DSM	Netherlands	NA	NA

NA, not available; PBSA, poly(butylene succinate adipate).

Babu *et al.*

Babu *et al. Progress in Biomaterials* 2013 2:8, doi:10.1186/2194-0517-2-8

Conventional processes for the production of 1,4-butanediol use fossil fuel feedstocks such as acetylene and formaldehyde. The bio-based process involves the use of glucose from renewable resources to produce succinic acid followed by a chemical reduction to produce butanediol. PBS is produced by transesterification, direct polymerization, and condensation polymerization reactions. PBS copolymers can be produced by adding a third monomer such as sebacic acid, adipic acid, and succinic acid which is also produced by renewable resources (Bechthold et al. 2008).

PBS is a semicrystalline polyester with a melting point higher than that of PLA. Its mechanical and thermal properties depend on the crystal structure and the degree of crystallinity (Nicolas et al.2011). PBS displays similar crystallization behavior and mechanical properties to those of polyolefin such as polyethylene. It has a good tensile and impact strength with moderate rigidity and hardness. The T_g is approximately $-32°C$, and the melting temperature is approximately 115°C. In comparison with PLA, PBS is tougher in nature but with a lower rigidity and Young's modulus. By changing the monomer composition, mechanical properties can be tuned to suit the required application (Liu et al. 2009a, b).

PBS and their blends have found commercial applications in agriculture, fishery, forestry, construction, and other industrial fields which are listed in Table 7. For example, PBS has been employed as mulch film, packaging, and flushable hygiene products and also used as a non-migrant plasticizer for polyvinyl chloride (PVC). In addition, it is used in foaming and food packaging application. The relatively poor mechanical flexibility of PBS limits the applications of 100% PBS-based products. However, this can be overcome by blending PBS with PLA or starch to improve the mechanical properties significantly, providing properties similar to that of polyolefin (Eslmai and Kamal 2013; Zhao et al. 2010).

Table 7: Applications of PBS and their blends

Polymer type	Applications	Reference
PBS/PLA blend	Packaging films, dishware, fibers, medical materials	Weraporn et al. 2011; Liu et al. 2009 a, b; Bhatia et al.2007; Lee and Wang 2006
PBS and blends	Drug encapsulation systems	Cornelia et al. 2011
PBS/starch	Barrier films	Jian-Bing et al. 2011
PBS and copolymers	Industrial applications	Jun and Bao-Hua 2010 a, b
PBS ionomers	Orthopedic applications	Jung et al. 2009

Babu *et al.*

Babu *et al. Progress in Biomaterials* 2013 2:8, doi:10.1186/2194-0517-2-8

Bio-polyethylene

Polyethylene (PE) is an important engineering polymer traditionally produced from fossil resources. PE is produced by polymerization of ethylene under pressure, temperature, in the presence of a catalyst. Traditionally, ethylene is produced through steam cracking of naphtha or heavy oils or ethanol dehydration. With increases in oil prices, microbial PE or green PE is now being manufactured from dehydration of ethanol produced by microbial fermentation. The concept of producing PE from bioethanol is not a particularly new one. In the 1980s, Braskem made bio-PE and bio-PVC from bioethanol. However, low oil prices and the limitations of the biotechnology processes made the technology unattractive at that time (de Guzman 2010).

Currently, bio-PE produced on an industrial scale from bioethanol is derived from sugarcane. Bioethanol is also derived from biorenewable feedstocks, including sugar beet, starch crops such as maize, wood, wheat, corn, and other plant wastes through microbial strain and biological fermentation process. In a typical process, extracted sugarcane juice with high sucrose content is anaerobically fermented to produce ethanol. At the end of the fermentation process, ethanol is distilled in order to remove water and to yield azeotropic mixture of hydrous ethanol. Ethanol is then dehydrated at high temperatures over a solid catalyst to produce ethylene and, subsequently, polyethylene (Guangwen et al. 2007; Luiz et al. 2010).

Bio-based polyethylene has exactly the same chemical, physical, and mechanical properties as petrochemical polyethylene. Braskem (Brazil) is the largest producer of bio-PE with 52% market share, and this is the first certified bio-PE in the world. Similarly, Braskem is developing other bio-based polymers such as bio-polyvinyl chloride, bio-polypropylene, and their copolymers with similar industrial technologies. The current Braskem bio-based PE grades are mainly targeted towards food packing, cosmetics, personal care, automotive parts, and toys. Dow Chemical (USA) in cooperation with Crystalsev is the second largest producer of bio-PE with 12% market share. Solvay (Belgium), another producer of bio-PE, has 10% share in the current market. However, Solvay is a leader in the production of bio-PVC with similar industrial technologies. China Petrochemical Corporation also plans to set up production facilities in China to produce bio-PE from bioethanol (Haung et al. 2008).

Bio-PE can replace all the applications of current fossil-based PE. It is widely used in engineering, agriculture, packaging, and many day-to-day commodity applications because of its low price and good performance. Table 8 shows applications of bio-PE in different fields where it can replace conventional PE.

Table 8: Application of bio-PE polymer and their blends

Table 8		
Application of bio-PE polymer and their blends		
Polymer type	Applications	Reference
Bio-PE	Plastics bags, milk and water bottles, food packaging films, toys	Vona et al. 1965; Aamer et al. 2008
Bio-PE and blends	Agricultural mulch films	Kasirajan and Ngouajio 2012

Babu *et al.*

Babu *et al. Progress in Biomaterials* 2013 2:8, doi:10.1186/2194-0517-2-8

Bio-Based Natural Polymers

This group consists of naturally occurring polymers such as cellulose, starch, chitin, and various polysaccharides and proteins. These materials and their derivatives offer a wide range of properties and applications. In this section, some of the natural bio-based polymers and their applications in various fields are discussed.

Starch

Starch is a unique bio-based polymer because it occurs in nature as discrete granules. Starch is the end product of photosynthesis in plants - a natural carbohydrate-based polymer that is abundantly available in nature from various sources including wheat, rice, corn, and potato. Essentially, starch consists of the linear polysaccharide amylose and the highly branched polysaccharide amylopectin. In particular, thermoplastic starch is of growing interest within the industry. The thermal and mechanical properties of starch can vary greatly and depend upon such factors as the amount of plasticizer present. The Tg varies between −50°C and 110°C, and the modulus is similar to polyolefins (Jane 1995). Several challenges exist in producing commercially viable starch plastics. Starch›s molecular structure is complex and partly nonlinear, leading to issues with ductility. Starch and starch thermoplastics suffer from the phenomenon of retrogradation - a natural increase in crystallinity over time, leading to increased brittleness. Plasticizers need to be found to create starch plastics with mechanical properties comparable to polyolefin-derived packaging. Plasticized starch blends and composites and/or chemical modifications may overcome these issues, creating biodegradable polymers with sufficient mechanical strength, flexibility, and water barrier properties for commercial packaging and consumer products (Maurizio et al. 2005).

Novamont is one of the leading companies in processing starch-based products (Li et al. 2009). The company produces various types of starch-based products using proprietary blend formulations. There are other companies around the world producing starch-based products in a similar scale for various applications, which are listed in Table 9 (Doug 2010; Ravenstijn 2010).

Table 9: Global suppliers of starch-based products

Company	Location	Brand name	Production/planned capacity (kton/year)
Novamont	Italy	Mater-Bi	120
Japan Corn Starch	Japan	Ever Corn	NA
Biotec	Germany	Bioplast	NA
Rodenberg	Netherlands	Solanyl	50
BIOP	Germany	Biopar	5
Plantic	Australia	Plantic	7.5
Wuhan Huali Environment Protection Sci. & Tech	China	PSM	15
Biograde	China	Cardia	3
PSM	USA	Plaststarch	NA
Livan	Canada	Livan	10

Babu *et al.*

Babu *et al. Progress in Biomaterials* 2013 2:8, doi:10.1186/2194-0517-2-8

Applications of thermoplastic starch polymers include films, such as for shopping, bread, and fishing bait bags, overwraps, flushable sanitary product, packing materials, and special mulch films. Potential future applications could include foam loose-fill packaging and injection-molded products such as 'take-away' food containers. Starch and modified starches have a broad range of applications both in the food and non-food sectors. In Europe in 2002, the total consumption of starch and starch derivatives was approximately 7.9 million tons, of which 54% was used for food applications and 46% in non-food applications (Frost & Sullivan report 2009).

The largest users of starch in the European Union (30%) are the paper, cardboard, and corrugating industries (Frost & Sullivan report 2009). Other important fields of starch application are textiles, cosmetics, pharmaceuticals, construction, and paints, which are listed in Table 10. In the medium and long term, starch will play an increasing role in the field of 'renewable raw materials' for the production of biodegradable plastics, packaging material, and molded products.

Table 10: Application of starch and their blends in various fields

Polymer type	Applications	Reference
Starch	Orthopedic implant devices as bone fillers	Ashammakhi and Rokkanen 1997
Starch/ethylene vinyl alcohol/ HA starch/polycaprolactone blends	Bone replacement/fixation implants, orthopedic applications	Mainil et al. 1997; Mendes et al. 2001; Marques and Reis 2005
Starch/cellulose acetate blends with methylmethacrylate and acrylic acid	Bone cements	Espigares et al. 2002
Modified starch	Food applications	Jaspreet et al. 2007; Fuentes et al. 2010
Starch derivatives	Drug delivery	Asha and Martins 2012
Thermoplastic starch	Packaging, containers, mulch films, textile sizing agents, adhesives	Zhao et al. 2008; Maurizio et al. 2005; Ozdemir and Floros 2004; Dave et al. 1999; Guo et al. 2005; Kumbar et al. 2001; Li et al. 2011

Babu *et al.*

Babu *et al. Progress in Biomaterials* 2013 2:8, doi:10.1186/2194-0517-2-8

Cellulose

Cellulose is the predominant constituent in cell walls of all plants. Cellulose is a complex polysaccharide with crystalline morphology. Cellulose differs from starch where glucose units are linked by β-1,4-glycosidic bonds, whereas the bonds in starch are predominantly α-1,4 linkages. The most important raw material sources for the production of cellulosic plastics are cotton fibers and wood. Plant fiber is dissolved in alkali and carbon disulfide to create viscose, which is then reconverted to cellulose in cellophane form following a sulfuric acid and sodium sulfate bath. There are currently two processes used to separate cellulose from the other wood constituents (Yan et al.2009). These methods, sulfite and pre-hydrolysis kraft pulping, use high pressure and chemicals to separate cellulose from lignin and hemicellulose, attaining greater than 97% cellulose purity. The main derivatives of cellulose for industrial purposes are cellulose acetate, cellulose esters (molding, extrusion, and films), and regenerated cellulose for fibers.

Cellulose is a hard polymer and has a high tensile strength of 62 to 500 MPa and elongation of 4% (Bisanda and Ansell 1992; Eichhorn et al. 2001). In order to overcome the inherent processing problems of cellulose, it is necessary to modify, plasticize, and blend with other polymers. The mechanical and thermal properties vary from blend to blend depending on the composition. The *Tg* of cellulosic derivatives ranged between 53°C and 180°C (Picker and Hoag 2002).

Eastman Chemical is a major producer of cellulosic polymers. FKuR launched a biopolymer business in the year 2000 and has a capacity of 2,800 metric ton/year of various cellulosic compounds for different applications (Doug 2010). The major produ

Table 11: Global suppliers of cellulosic products

Company	Location	Brand name
Innovia films	UK	Nature Flex
Eastman Chemical	USA	Tenite

FKuR	Germany	Biograde
Sateri	China	Sateri

Babu *et al.*

Babu *et al. Progress in Biomaterials* 2013 2:8, doi:10.1186/2194-0517-2-8

There are three main groups of cellulosic polymers that are produced by chemical modification of cellulose for various applications. Cellulose esters, namely cellulose nitrate and cellulose acetate, are mainly developed for film and fiber applications. Cellulose ethers, such as carboxymethyl cellulose and hydroxyethyl cellulose, are widely used in construction, food, personal care, pharmaceuticals, paint, and other pharmaceutical applications (Kamel et al. 2008). Finally, regenerated cellulose is the largest bio-based polymer produced globally for fiber and film applications. Regenerated cellulose fibers are used in textiles, hygienic disposables, and home furnishing fabrics because of its thermal stability and modulus (Kevin et al. 2001).

Chemically pure cellulose can be produced using a certain type of bacteria. Bacterial cellulose is characterized by its purity and high strength. It can be used to produce articles with relatively high strength. Currently, applications for bacterial cellulose outside food and biomedical fields are rather limited because of its high price. The other applications include acoustic diaphragms, mining, paints, oil gas recovery, and adhesives. However, the low yields and high costs of bacterial cellulose represent barriers to large-scale industrial applications (Prashant et al. 2009). Table 12 summarizes the applications of cellulose and their compounds in different fields.

Table 12: Application of cellulose and their compounds in various fields

Table 12		
Application of cellulose and their compounds in various fields		
Polymer type	**Applications**	**Reference**
Cellulose esters	Membranes for separation	Kumano and Fujiwara 2008
Carboxylated methyl cellulose	Drug formulations, as binder for drugs, film-coating agent for drugs, ointment base	Chambin et al. 2004; Obae and Imada 1999; Westermark et al. 1999; Hirosawa et al. 2000

Cellulose acetate fibers	Wound dressings	Orawan et al. 2008; Abdelrahman and Newton 2011
Hydroxyethyl cellulose	Spray for clothes polluted with pollen	Hori et al. 2005
Modified celluloses, cellulose whiskers, microfibrous cellulose	Barrier films, water preservation in food packing	Amit and Ragauskas 2009
Cellulose nanofibers	Textile applications	Zeeshan et al. 2013
Cellulose particles	Chromatographic applications, chiral separations	Levison 1993; Arshady 1991a, b

Babu *et al.*

Babu *et al. Progress in Biomaterials* 2013 2:8, doi:10.1186/2194-0517-2-8

Chitin and Chitosan

Chitin and chitosan are the most abundant natural amino polysaccharide and valuable bio-based natural polymers derived from shells of prawns and crabs. Currently, chitin and chitosan are produced commercially by chemical extraction process from crab, shrimp, and prawn wastes (Roberts 1997). The chemical extraction of chitin is quite an aggressive process based on demineralization by acid and deproteination by the action of alkali followed by deacetylated into chitosan (Roberts 1997). Chitin can also be produced by using enzyme hydrolysis or fermentation process, but these processes are not economically feasible on an industrial scale (Win and Stevens2001). Currently, there are few industrial-scale plants of chitin and chitosan worldwide located in the USA, Canada, Scandinavia, and Asia (Ravi Kumar 2000).

Chitosan displays interesting characteristics including biodegradability, biocompatibility, chemical inertness, high mechanical strength, good film-forming properties, and low cost (Marguerite 2006; Virginia et al. 2011; Liu et al. 2012). Chitosan is being used in a vast array of widely varying products and applications ranging from pharmaceutical and cosmetic products to water treatment and plant protection. For each application, different properties of chitosan are required, which changes with the degree of acetylation and molecular weight. Chitosan is compatible with many biologically active components incorporated in cosmetic product composition (Ravi Kumar 2000). Due to its low toxicity, biocompatibility, and bioactivity, chitosan has become a very attractive material in such diverse applications as biomaterials in medical devices and as a pharmaceutical ingredient (Bae and Moo-Moo 2010; Ramya et al. 2012). Chitosan has application in shampoos, rinses, and permanent hair-coloring agents. Chitosan and its derivatives also have applications in the skin care industry. Chitosan can function as a moisturizer for the skin, and because of its lower costs, it might compete with hyaluronic acid in this application (Bansal et al. 2011; Valerie and Vinod 1998; Hafdani and Sadeghinia 2011).

Pullulan

Pullulan is a linear water-soluble polysaccharide mainly consisting of maltotriose units connected by α-1,6 glycosidic units. Pullulan was first reported by Bauer (1938) and is obtained from the fermentation broth of *Aureobasidium pullulans*. Pullulan is produced by a simple fermentation process using a number of feedstocks containing simple sugars (Bernier 1958; Catley 1971; Sena et al. 2006). Pullulan can be chemically modified to produce a polymer that is either less soluble or completely insoluble in water. The unique properties of this polysaccharide are due to its characteristic glycosidic linking. Pullulan is easily chemically modified to reduce the water solubility or to develop pH sensitivity, by introducing functional reactive groups, etc. Due to its high water solubility and low viscosity, pullulan has numerous commercial applications including use as a food additive, a flocculant, a blood plasma substitute, an adhesive, and a film (Zajic and LeDuy 1973; Singh et al. 2008; Cheng et al. 2011). Pullulan can be formed into molding articles which can resemble conventional polymers such as polystyrene in their transparency, strength, and toughness (Leathers 2003).

Pullulan is extensively used in the food industry. It is a slow-digesting macromolecule which is tasteless as well as odorless, hence its application as a low-calorie food additive providing bulk and texture. Pullulan possesses oxygen barrier property and good moisture retention, and also, it inhibits fungal growth. These properties make it an excellent material for food preservation, and it is used extensively in the food industry (Conca and Yang 1993). In recent years, pullulan has also been studied for biomedical applications in various aspects, including targeted drug and gene delivery, tissue engineering, wound healing, and even in diagnostic imaging medium (Rekha and

Chrndra 2007). Other emerging markets for pullulan include oral care products (Barkalow et al.2002) and formulations of capsules for dietary supplements and pharmaceuticals (Leathers 2003), leading to increased demand for this unique biopolymer.

Collagen and Gelatin

Collagen is the major insoluble fibrous protein in the extracellular matrix and in connective tissue. In fact, it is the single most abundant protein in the animal kingdom. There are at least 27 types of collagens, and the structures all serve the same purpose: to help tissues withstand stretching. The most abundant sources of collagen are pig skin, bovine hide, and pork and cattle bones. However, the industrial use of collagen is obtained from nonmammalian species (Gomez-Guille et al. 2011). Gelatin is obtained through the hydrolysis of collagen. The degree of conversion of collagen into gelatin depends on the pretreatment, function of temperature, pH, and extraction time (Johnston-Banks 1990).

Collagen is one of the most useful biomaterials due to its biocompatibility, biodegradability, and weak antigenicity (Maeda et al. 1999). The main application of collagen films in ophthalmology is as drug delivery systems for slow release of incorporated drugs (Rubin et al. 1973). It was also used for tissue engineering including skin replacement, bone substitutes, and artificial blood vessels and valves (Lee et al. 2001).

The classical food, photographic, cosmetic, and pharmaceutical applications of gelatin is based mainly on its gel-forming properties. Recently in the food industry, an increasing number of new applications have been found for gelatin in products in line with the growing trend to replace synthetic agents with more natural ones (Gomez-Guille et al. 2011). These include emulsifiers, foaming agents, colloid stabilizers, biodegradable film-forming materials, and microencapsulating agents.

Alginates

Alginate is a linear polysaccharide that is abundant in nature as it is synthesized by brown seaweeds and by soil bacteria (Draget et al. 1997). Sodium alginate is the most commonly used alginate form in the industry since it is the first by-product of algal purification (Draget 2000). Sodium alginate consists of α-*l*-guluronic acid residues (G blocks) and β-*d*-mannuronic acid residues (M blocks), as well as segments of alternating guluronic and mannuronic acids.

Although alginates are a heterogeneous family of polymers with varying content of G and M blocks depending on the source of extraction, alginates with high G content have far more industrial importance (Siddhesh and Edgar 2012). The acid or alkali treatment processes used to make sodium alginate from brown seaweeds are relatively simple. The difficulties in processing arise mainly from the separation of sodium alginate from slimy residues (Black and Woodward 1954). It is estimated that the annual production of alginates is approximately 38,000 tons worldwide (Helgerud et al. 2009).

Alginates have various industrial uses as viscosifiers, stabilizers, and gel-forming, film-forming, or water-binding agents (Helga and Svein 1998). These applications range from textile printing and manufacturing of ceramics to production of welding rods and water treatment (Teli and Chiplunkar1986; Qin et al. 2007; Xie et al. 2001). The polymer is soluble in cold water and forms thermostable gels. These properties are utilized in the food industry in products such as custard creams and restructured food. The polymer is also used as a stabilizer and thickener in a variety of beverages, ice creams, emulsions, and sauces (Iain et al. 2009).

Alginates are widely used as a gelling agent in pharmaceutical and food applications. Studies into their positive effects on human health have broadened recently with the recognition that they have a number of potentially beneficial physiological effects in the gastrointestinal tract (Peter et al.2011; Mandel et al. 2000). Alginate-containing wound dressings are commonly used, especially in making hydrophilic gels over wounds which can produce comfortable, localized hydrophilic environments in healing wounds (Onsoyen 1996). Alginates are used in controlled drug delivery, where the rate of drug release depends on the type and molecular weight of alginates used (Alexnader et al. 2006; Goh et al. 2012). Additionally, dental impressions made with alginates are easy to handle for both dentist and patient as they fast set at room temperature and are cost-effective (Onsoyen 1996). Recent studies show that alginates can be effective in treating obesity, and currently, various functional alginates are being evaluated in human clinical trials (Georg et al.2012).

Current Status and Future Trends

The use of bio-based feedstocks in the chemical sector is not a novel concept. They have been industrially feasible on a large scale for more than a decade. However, the price of oil was so cost-effective, and the development of oil-based products created so many opportunities that bio-based products were not prioritized at the time. Several factors, such as the limitations and uncertainty in supplies of fossil fuels, environmental considerations, and technological developments, accelerated the advancement of bio-based polymers and products. It took more than a century to evolve the fossil fuel-based chemical industry; however, the bio-based polymer industry is already catching up with fossil fuel-based chemical industry, which has augmented in the last 20 years. Thanks to advancements in white biotechnology, the production of bio-based polymers and other chemicals from renewable resources has become a reality. The first-generation technologies mainly focused on food resources such as corn, starch, rice, etc. to produce bio-based polymers. As the food-versus-fuel debate ascended, the focus of technologies diverted to cellulose-based feedstocks, focusing on waste from wood and paper, food industries, and even stems and leaves and solid municipal waste streams. More and more of these technologies are already in the pipeline to align with the abovementioned waste streams; however, it may take another 20 years to develop the full spectrum of chemicals based on these technologies (Michael et al. 2011).

Challenges that need to be addressed in the coming years include management of raw materials, performance of bio-based materials, and their cost for production. Economy of scale will be one of the main challenges for production of bio-based monomers and bio-based polymers from renewable sources. Building large-scale plants can be difficult due to the lack of experience in new technologies and estimation of supply/demand balance. In order to make these technologies economically viable, it is very important to develop (1) logistics for biomass feedstocks, (2) new manufacturing routes by replacing existing methods with high yields, (3) new microbial strains/enzymes, and (4) efficient downstream processing methods for recovery of bio-based products.

The current bio-based industry focus is mainly on making bio-versions of existing monomers and polymers. Performance of these products is well known, and it is relatively easy to replace the existing product with similar performance of bio-versions. All the polymers mentioned above often display similar properties of current fossil-based polymers. Recently, many efforts are seen towards introducing new bio-based polymers with higher performance and value. For example, Nature Works LLC has introduced new grades of PLA with higher thermal and mechanical properties. New PLA-tri block copolymers have been reported to behave like thermoplastic elastomer. Many developments are currently underway to develop various polyamides, polyesters, polyhydroxyaloknates, etc. with a high differentiation in their final properties for use in automotive, electronics, and biomedical applications.

The disadvantage of some of the new bio-based polymers is that they cannot be processed in all current processing equipment. There is vast knowledge on additive-based chemistry developed for improving the performance and processing of fossil fuel-based polymers, and this knowledge can be used to develop new additive chemistry to improve the performance and properties of bio-based polymers (Ray and Bousmina 2005). For bio-based polymers like PLA and PHA, additives are being developed to improve their performance, by blending with other polymers or making new copolymers. However, the additive market for bio-based polymers is still very small, which makes it difficult to justify major development efforts according to some key additive supplier companies.

The use of nanoparticles as additives to enhance polymer performance has long been established for petroleum-based polymers. Various nano-reinforcements currently being developed include carbon nanotubes, graphene, nanoclays, 2-D layered materials, and cellulose nanowhiskers. Combining these nanofillers with bio-based polymers could enhance a large number of physical properties, including barrier, flame resistance, thermal stability, solvent uptake, and rate of biodegradability, relative to unmodified polymer resin. These improvements are generally attained at low filler content, and this nano-reinforcement is a very attractive route to generate new functional biomaterials for various applications.

Even though new bio-based polymers are produced on an industrial scale, there are still several factors which need to be determined for the long-term viability of bio-based polymers. It is expected that there will be feedstock competition as global demand for food and energy increases over time. Currently, renewable feedstocks used for manufacturing bio-based monomers and polymers often compete with requirements for food-based products. The expansion of first-generation bio-based fuel production will place unsustainable demands on biomass resources and is as much a threat to the sustainability of biochemical and biopolymer production as it is to food production (Michael et al. 2011). Indeed the European commission has altered its targets downwards for first-generation biofuels since October 2012, indicating its preference for non-food sources of sugar for biofuel production (EurActiv.com 2012). Several initiatives are underway to use cellulose-based feedstocks for the production of usable sugars for biofuels, biochemicals, and biopolymers (Jong et al. 2010).

CONCLUSIONS

Bio-based polymers are closer to the reality of replacing conventional polymers than ever before. Nowadays, bio-based polymers are commonly found in many applications from commodity to hi-tech applications due to advancement in biotechnologies and public awareness. However, despite these advancements, there are still some drawbacks which prevent the wider commercialization of bio-based polymers in many applications. This is mainly due to performance and price when compared with their conventional counterparts, which remains a significant challenge for bio-based polymers.

AUTHORS' CONTRIBUTIONS

RPB contributed in writing the whole manuscript. KOC contributed in providing the information on applications and policy information of bio-based polymers. SR contributed in providing the outline for the manuscript. All authors read and approved the final manuscript

ACKNOWLEDGMENTS

RPB would like to acknowledge the financial support from the Environmental Protection Agency, Ireland, under grant no. 2008-ET-LS-1-S2.

REFERENCES

1. Aamer AS, Fariha H, Abdul H, Safia A (2008) Biological degradation of plastics: a comprehensive review. Biotechnol Adv 26:246-265

2. Abdelrahman T, Newton H (2011) Wound dressings: principles and practice. Surgery 29:491-495

3. Alexnader DA, Kong HJ, Mooney DJ (2006) Alginate hydrogels as biomaterials. Macromolecular Biosciences 6:623-633

4. Amass W, Amass A, Tighe B (1998) A review of biodegradable polymers: uses, current developments in the synthesis and characterization of biodegradable polyesters, blends of biodegradable polymers and recent advances in biodegradation studies. Polymer International 47:89-144

5. Amit S, Ragauskas AJ (2009) Water transmission barrier properties of biodegradable films based on cellulosic whiskers and xylan. Carbohydr Polym 78(2):357-360

6. Arshady R (1991) Beaded polymer supports and gels: 2. Physicochemical criteria and functionalization. J Chromatogr 586:199-219

7. Arshady R (1991) Beaded polymer supports and gels: 1. Manufacturing techniques. J Chromatogr 586:181-197

8. Asha R, Martins E (2012) Recent applications of starch derivatives in nanodrug delivery. Carbohydr Polym 87(2):987-994

9. Ashammakhi N, Rokkanen P (1997) Absorbable polyglycolide devices in trauma and bone surgery. Biomaterials 18(1):3-9

10. Avinc A, Akbar K (2009) Overview of poly (lactic acid) fibres. Part I: production, properties, performance, environmental impact, and end-use applications of poly (lactic acid) fibres. Fiber Chemistry 41(6):391-401

11. Bae KP, Moo-Moo K (2010) Applications of chitin and its derivatives in biological medicine. Int J Mol Sci 11:5152-5164

12. Bansal V, Pramod KS, Nitin S, Omprakask P, Malviya R (2011) Applications of chitosan and chitosan derivatives for drug delivery. Adva Biol Res 5:28-37

13. Barkalow DG, Chapedelaine AH, Dzija MJ (2002) Improved pullulan free edible film compositions and methods of making same. PCT International Application WO 02/43657, US 01/43397, 21 Nov.

14. Bauer R (1938) Physiology of Dematium pullulans de Bary. Zentralbl Bacteriol Parasitenkd Infektionskr Hyg Abt2 98:133-167

15. Bayram C, Denbas EB (2008) Preparation and characterization of triamcinolone acetonide-loaded poly(3-hydroxybutyrate-co-3-hydroxyhexanoate) (PHBHx) microspheres. J Bioactive and Compatible Polymer 23:334-347

16. Bechthold I, Bretz K, Kabasci S, Kopitzky R, Springer A (2008) Succinic acid: a new platform chemical from biobased polymers from renewable resources. Chemical Engg Technol 31:647-654

17. Bernier B (1958) The production of polysaccharides by fungi active in the decomposition of wood and forest litter. Can J Microbiol 4:195-204

18. Bhatia A, Gupta RK, Bhattacharaya SN, Choi HJ (2007) Compatibility of biodegradable PLA and PBS blends for packaging applications. Korea Aust Rheol J 19:125-131

19. Bisanda ETN, Ansell MP (1992) Properties of sisal-CNSL composites. J Mater Sci 27:1690-1700

20. Black WAP, Woodward FN (1954) Alginates from common British brown marine algae. In Natural plant hydrocolloids. Adv Chem Ser Am Chem Soc 11:83-91

21. Catley BJ (1971) Utilization of carbon sources by Pullularia pullulans for the elaboration of extracellular polysaccharides. Appl Microbiol 22:641-649

22. Chambin DC, Debray C, Rochat-Gonthier MH, Le MM, Pourcelot M (2004) Effects of different cellulose derivatives on drug release mechanism studied at a pre-formulation stage. J Controll Release 95(1):101-108

23. Chandy T, Das GS, Rao GH (2000) 5-Fluorouracil-loaded chitosan coated polylactic acid microspheres as biodegradable drug carriers for cerebral tumours. J Microencapsul 5:625-631

24. Chen GQ, Qiong W (2005) The application of polyhydroxyalkanoates as tissue engineering materials. Biomaterials 26:6565-6578

25. Chen GQ, Martin KP (2012) Plastics derived from biological sources: present and future: a technical and environmental review. Chem Rev 112:2082-2099

26. Chen QZ, Harding SE, Ali NN, Lyon AR, Boccaccini AR (2008) Biomaterials in cardiac tissue engineering: ten years of research survey. Materials Sci Eng: Reports 59:1-37

27. Cheng KC, Demirci A, Catchmark JM (2011) Pullulan: biosynthesis, production, and applications. Appl Microbiol Biotechnol 92:29-44

28. Cheng Y, Deng S, Chen P, Ruan R (2009) Polylactic acid (PLA) synthesis and modifications: a review. Front Chem China 4:259-264

29. Clarinval AM, Halleux J (2005) Classification of biodegradable polymers. In: Smith R (ed) Biodegradable polymers for industrial applications, Woodhead, Cambridge.

30. Conca KR, Yang TCS (1993) Edible food barrier coatings. In: Ching C, Kaplan DL, Thomas EL (eds) Biodegradable polymers and packaging, Technomic, Lancaster. pp 357-369

31. Cornelia TB, Erkan TB, Elisabete DP, Rui LR, Nuno MN (2011) Performance of biodegradable microcapsules of poly(butylene succinate), poly(butylene succinate-co-adipate) and poly(butylene terephthalate-co-adipate) as drug encapsulation systems. Colloids Surf B Biointerfaces 84:498-507

32. Coutu DL, Yousefi AM, Galipeau J (2009) Three-dimensional porous scaffolds at the crossroads of tissue engineering and cell-based gene therapy. J Cell Biochem 108:537-546

33. Dave AM, Mehta MH, Aminabhavi TM, Kulkarni AR, Soppimath KS (1999) A review on controlled release of nitrogen fertilizers through polymeric membrane devices. Polymer- Plastics Technol Eng 38:675-711

34.   de Roo G, Kellerhals MB, Ren Q, Witholt B, Kessler B (2002) Production of chiral R-3-hydroxyalkanoic acids and R-3-hydroxyalkanoic acid methylesters via hydrolytic degradation of polyhydroxyalkanoate synthesized by pseudomonads. Biotechnol Bioeng 77:717-722

35.   de Guzman D (2010) Bioplastic development increases with new applications. http://www.icis.com/Articles/2010/10/25/9402443/bioplastic-development-increases-with-new-applications.html . Accessed October 2010

36.   Dorozhkin SV (2009) Calcium orthophosphate-based biocomposites and hybrid biomaterials. J Mater Sci 44:2343-2387

37.   Doug S (2010) Bioplastics: technologies and global markets. BCC research reports PLS050A. http://www.bccresearch.com/report/bioplastics-technologies-markets-pls050a.html

38.   Draget KI (2000) Alginates. In: Philips O, Williams A (eds) Handbook of hydrocolloids, Woodhead, Philadelphia. p 379

39.   Draget KI, Skjåk-Braek G, Smidsrød O (1997) Alginate based new materials. Int J Biol Macromol 21:47-55

40.   DSM press release (2008) DSM invests in development of bio-based materials. http://www.observatorioplastico.com/detalle_noticia.php?no_id=73274&seccion=mercado&id_categoria=80002 . Accessed March 2008

41.   Eichhorn SJ, Baillie CA, Zaferiropouls N, Mwaikambo LY, Ansell MP, Dufresne A, Entwistle KM, Herrera-Franco PJ, Escamilla GC, Groom L, Hughes M, Hill C, Rials TG, Wild PM (2001) Review: current international research into cellulosic fibres and composites. J Material Sc 36:2107-2131

42.   Erwin TH, David AG, Jeffrey JK, Robert JW, Ryan PO (2007) The eco-profiles for current and near-future NatureWorks® polylactide (PLA) production. Industrial Biotechnology 3:58-81

43.   Eslmai H, Kamal RM (2013) Elongational rheology of biodegradable poly(lactic acid)/poly[(butylene succinate)-co-adipate] binary blends and poly(lactic acid)/poly[(butylene succinate)-co-adipate]/clay ternary nanocomposites. J Appl Polym Sci 127:2290-2306

44. Espigares I, Elvira C, Mano JF, Vlazquez B, Roman JS, Reis RL (2002) New partially degradable and bioactive acrylic bone cements based on starch blends and ceramic fillers. Biomaterials 23(8):1883-1895

45. EurActiv.com (2012) EU calls time on first-generation biofuels. http://www.euractiv.com/climate-environment/eu-signals-generation-biofuels-news-515496. Accessed Oct 2012

46. Frost & Sullivan report (2009) Global bio-based plastic market, M4AI-39. Chapter 5. http://www.frost.com/sublib/display-report.do?ctxixpLink=FcmCtx9&searchQuery=Global+bio-based+plastic+market%2C+2009&bdata=aHR0cDovL3d3dy5mcm9zdC5jb20vc3JjaC9jYXRhbG9nLXNlYXJjaC5kbz9wdWJsaWNhdGlvblYXJzPTIwMDkmcXVlcnlUZXh0OPUdsb2JhbCtiaWFW8tYmFzZWQrcGxhc3RpYyttYXJrZXQlMkMrMjAwOSZmaWx0ZXJzPTIl0Y2g9ZmFsc2UmcGFnZVNpemU9MTAmTJAfkBTZWFyY2gggUmVzdWx0c0B%2BQDEzNjMxMDQzNjA1NjA%3D&ctxixpLabel=FcmCtx10&id=M4A1-01-00-00-00. Accessed 22 Dec 2009

47. Freier T, Kunze C, Nischan C (2002) In vitro and in vivo degradation studies for development of a biodegradable patch based on poly(3-hydroxybutyrate). Biomaterials 23:2649-2657

48. Fuentes Z, Riquelme MJN, Sánchez-Zapata E, Pérez JAÁ (2010) Resistant starch as functional ingredient: a review. Food Res Int 43:931-942

49. Garlotta D (2001) A literature review of poly (lactic acid). J Polyms and the Envir 9(2):63-84

50. Georg JM, Kristensen M, Astrup A (2012) Effect of alginate supplementation on weight loss in obese subjects completing a 12-week energy restricted diet: a randomized controlled trail. Am J Clin Nutr 96:5-13

51. Goh GH, Heng PWS, Chan LW (2012) Alginates as a useful natural polymer for microencapsulation and therapeutic applications. Carbohydr Polym 88:1-12

52. Gomez-Guille MC, Gimenez B, Lopez CME, Montero MP (2011) Functional bioactive properties of collagen and gelatin from alternative sources: a review. Food Hydrocolloids 25:1813-1827

53. Guangwen C, Shulian L, Fengjun J, Quan Y (2007) Catalytic dehydration of bioethanol to ethylene over $TiO_2/\gamma-Al_2O_3$ catalyst in microchannel reactors. Catal today 125:111-119

54. Guo M, Liu M, Zhan F, Wu L (2005) Preparation and properties of a slow-release membrane-encapsulated urea fertilizer with superabsorbent and moisture preservation. Ind Eng Chem Res 44:4206-4211

55. Gupta B, Revagade N, Hilborn J (2007) Poly(lactic acid) fiber: an overview. Prog Polym Sci 34:455-482

56. Hafdani FN, Sadeghinia N (2011) A review on applications of chitosan as a natural antimicrobial. World Academy of Sci Engg Technol 50:252-256

57. Hanggi JU (1995) Requirements on bacterial polyesters as future substitute for conventional plastics for consumer goods. FEMS Microbioly Rev 16:213-220

58. Haung YM, Li H, Huang XJ, Hu YC, Hu Y (2008) Advances of bio-ethylene. Chin J Bioprocess Eng 6:1-6

59. Helga E, Svein V (1998) Biosynthesis and applications of alginates. Polym Degradation and Stability 59:85-91

60. Helgerud T, Gaserød O, Fjreide T, Andresen PO, Larsen CK (2009) Alginates. In: Imeson A (ed) Food stabilisers, thickeners and gelling agents, Wiley Blackwell, Oxford. pp 50-72

61. Hirosawa E, Danjo K, Sunada H (2000) Influence of granulating method on physical and mechanical properties, compression behavior, and compactibility of lactose and microcrystalline cellulose granules. Drug Dev Ind Pharm 26:583-593

62. Hori K, Nojiri H, Nonomura M, Okuda F, Yanagida H (2005) Allergen inactivator. US Patent 197319, 20 Nov 2005

63. Huan Z, Joseph GL, Sarit BB (2012) Fabrication aspects of PLA-CaP/PLGA-CaP composites for orthopedic applications: a review. Acta Biomater 8(6):1999-2016

64. Iain AB, Seal CJ, Wilcox M, Dettmar PW, Pearson PJ (2009) Applications of alginates in food. In: Brend HAR (ed) Alginates: biology and applications. Microbiology monographs 13, Springer, Hiedelberg. pp 211-228

65. Jane J (1995) Starch properties, modifications and applications. J Macromolecular Sci 32:751-757

66. Jaspreet S, Lovedeep K, McCarthy OJ (2007) Factors influencing the physico-chemical, morphological, thermal and rheological properties of some chemically modified starches for food applications—a review. Food Hydrocolloids 21:1-22

67. Jayanth P, Vinod L (2012) Biodegradable nanoparticles for drug and gene delivery to cells and tissue. Adv Drug Deliv Rev 64:61-71

68. Jeevitha D, Kanchana A (2013) Chitosan/PLA nanoparticles as a novel carrier for the delivery of anthraquinone: synthesis, characterization and *in vitro* cytotoxicity evaluation. Colloids Surf B Biointerfaces 101(1):126-134

69. Jian-Bing Z, Ling J, Yi-Dong L, Madhusudhan S, Tao L, Yu-Zhong W (2011) Bio-based blends of starch and poly(butylene succinate) with improved miscibility, mechanical properties, and reduced water absorption. Carbohydr Polym 83:762-768

70. Johnston-Banks FA (1990) Gelatin. In: Harris P (ed) Food gels, Elsevier, London. pp 233-289

71. Jong ED, Higson A, Walsh P, Maria W (2010) Bio-based chemicals: value added products from biorefineries. IEA Bioenergy Task 42 Biorefinery. 1-34 http://www.iea-bioenergy.task42 biorefineries.com/publications/reports/?eID=dam_frontend_push&docID=2051. Accessed 15 Feb 2012

72. Jun X, Bao-Hua G (2010) Microbial succinic acid, its polymer poly(butylene succinate), and applications. Microbiology Monographs 14:347-388

73. Jun X, Bao-Hua G (2010) Poly(butylene succinate) and its copolymers: research, development and industrialization. Biotechnol J 5:1149-1163

74. Jung S, Lim E, Jong HK (2009) New application of poly(butylene succinate) (PBS) based ionomer as biopolymer: a role of ion group for hydroxyapatite (HAp) crystal formation. J Mater Sci 44:6398-6403

75. Jung-Ju K, Guang-Zhen J, Hye-Sun Y, Seong-Jun C, Hae-Won K, Ivan BW (2012) Providing osteogenesis conditions to mesenchymal stem cells using bioactive nanocomposite bone scaffolds. Mater Sci Eng C 32:2545-2551

76. Kamel S, Ali N, Jahangir K, Shah SM, El-Gendy (2008) Pharmaceutical significance of cellulose: a review. Express polymer Letters 2:758-778

77. Kasirajan S, Ngouajio M (2012) Polyethylene and biodegradable mulches for agricultural applications: a review. Agronomy Sustainable Dev 32(2):501-529

78. Kathiraser Y, Aroua MK, Ramachandran KB, Tan IKP (2007) Chemical characterization of medium-chain-length polyhydroxyalkanoates (PHAs) recovered by enzymatic treatment and ultrafiltration. J Chem Tech Biotech 82:847-855

79. Kellomaki M, Niiranen H, Puumanen K, Ashammakhi N, Waris T, Tormala P (2000) Bioabsorbable scaffolds for guided bone regeneration and generation. Biomaterials 21:2495-2505

80. Kevin JE, Charles MB, John DS, Paul AR, Brian DS, Michael CS, Debra T (2001) Advances in cellulose eater performance and applications. Progress in Polymer Sci 26:1605-1688

81. Kumano A, Fujiwara N (2008) Cellulose triacetate membranes for reverse osmosis. In: Normam AGF, Li N, Winston Ho WS, Matsuura T (eds) Advanced membrane technology and application, Wiley, New Jersey. pp 21-43

82. Kumbar SG, Kulkarni AR, Dave AM, Aminabha TM (2001) Encapsulation efficiency and release kinetics of solid and liquid pesticides through urea formaldehyde cross-linked starch, guar gum, and starch + guar gum matrices. J Appli Polym Sci 82:2863-2866

83. Kunze C, Edgar Bernd H, Androsch R (2006) *In vitro* and *in vivo* studies on blends of isotactic and atactic poly (3-hydroxybutyrate) for development of a dura substitute material. Biomaterials 27:192-201

84. Leathers TD (2003) Biotechnological production and applications of pullulan. Appl Microbiol Biotechnol 62:468-473

85. Lee SH, Wang S (2006) Biodegradable polymers/bamboo fiber composite with bio-based coupling agent. Compos Part A 37:80-91

86. Lee HC, Anuj S, Lee Y (2001) Biomedical applications of collagen. International J of Pharmaceutics 221:1-22

87. Lemoigne M (1923) Production d'acide β-oxybutyrique par certaines bact'eries du groupe du Bacillus subtilis. CR. Hebd. Seances Acad. Sci 176:1761

88. Levison PR (1993) Cellulosics as ion-exchange materials. In: Kennedy JF, Phillips GO, Williams PA (eds) Cellulosics: materials for selective separations and other technologies, Ellis Horwood, Chichester. pp 25-36

89. Li G, Yong H, Chen C (2011) Discussion on application prospect of starch-based adhesives on architectural gel materials. Adv Materials Res 250:800-803

90. Li S, Juliane H, Martin KP (2009) Product overview and market projection of emerging biobased products. PRo-BIP 1:1-245

91. Liu L, Yu J, Cheng L, Qu W (2009) Mechanical properties of poly(butylene succinate) (PBS) biocomposites reinforced with surface modified jute fibre. Composites Part A: Appl Sci Manufacturing 40:669-674

92. Liu LF, Yu JY, Cheng LD, Yang XJ (2009) Biodegradability of PBS composite reinforced with jute. Polym Degrade Stab 94:90-94

93. Liu M, Zhang Y, Wu C, Xiong S, Zhou C (2012) Chitosan/halloysite nanotubes bionanocomposites: structure, mechanical properties and biocompatibility. Int J Biological Macromol 51:566-575

94. Luc A, Eric P (2012) Biodegradable polymers. In: Environmental silicate nano-biocomposites. Green energy and technology. Springer, Hiedelberg. pp 13-39

95. Luiz A, De Castro R, Morschbacker (2010) A method for the production of one or more olefins, an olefin, and a polymer. US 2010/0069691A1, 18 Mar 2010

96. Maeda M, Tani S, Sano A, Fujioka K (1999) Microstructure and release characteristics of the minipellet, a collagen based drug delivery system for controlled release of protein drugs. J Controlled Rel 62:313-324

97. Mainil V, Rahn B, Gogolewski S (1997) Long-term *in vivo* degradation and bone reaction to various polylactides: 1. One-year results. Biomaterials 18:257-266

98. McChalicher CW, Srienc F (2007) Investigating the structure–property relationship of bacterial PHA block copolymers. J Biotechnology 132:296-302

99. Ravi Kumar MNV (2000) A review of chitin and chitosan applications. Reactive Functional Polym 46:1-27

100. Majid J, Elmira AT, Muhammad I, Muriel J, St'ephane D (2010) Poly-lactic acid: production, applications, nanocomposites, and release studies. Comprehensive Rev Food Sci Safety 9(5):552-571

101. Mandel KG, Daggy BP, Brodie DA, Jacoby HI (2000) Review article: alginate-raft formulations in the treatment of heartburn and acid reflux. Aliment Pharmacol Ther 14:669-690

102. Marguerite R (2006) Chitin and chitosan: properties and applications. Progress in Polym Sci 31:603-632

103. Marques AP, Reis RL (2005) Hydroxyapatite reinforcement of different starch-based polymers affect osteoblast-like cells adhesion/spreading and proliferation. Mater Sci Engg 25(2):215-229

104. Maurizio A, Jan JDV, Maria EE, Sabine F, Paolo V, Maria GV (2005) Biodegradable starch/clay nanocomposite films for food packaging applications. Food Chem 93(3):467-474

105. Mendes RL, Reis YP, Bovell AM, Cunha CA, Blitterswijk V, de Bruijn JD (2001) Biocompatibility testing of novel starch-based materials with potential application in orthopedic surgery: a preliminary study. Biomaterials 22:2057-2064

106. Michael C, Dirk C, Harald K, Jan R, Joachim V (2011) Policy paper on bio-based economy in the EU: level playing field for bio-based chemistry and materials. www.bio-based.eu/policy/en . Accessed December 2012

107. Nagarwal RC, Singh PN, Kant S, Maiti P, Pandit JK (2010) Chitosan coated PLA nanoparticles for ophthalmic delivery: characterization, in-vitro and in-vivo study in rabbit eye. J Biomed Nanotechnol 6:648-656

108. Nicolas J, Floriane F, Francoise F, Alan R, Jean PP, Patrick F, Rene SL (2011) Synthesis and properties of poly(butylene succinate): efficiency of different transesterfication catalysts. J Polym Sci Part A: Polym Chem 49:5301-5312

109. Obae HI, Imada K (1999) Morphological effect of microcrystalline cellulose particles on tablet tensile strength. Int J Pharm 182:155-164

110. Onsoyen E (1996) Commercial applications of alginates. Carbohydrates in Europe 14:26-31

111. Orawan S, Uracha R, Pitt S (2008) Electrospun cellulose acetate fiber mats containing asiaticoside or Centella asiatica crude extract and the release characteristics of asiaticoside. Polymer 49(19):4239-4247

112. Ozdemir M, Floros JD (2004) Active food packaging technologies. Crit Rev Food Sci Nutr 44:185-193

113. Papenburg BJ, Liu J, Higuera G, Barradas AMC, Boer J, Blitterswijk VCA, Wessling M, Stamatialis D (2009) Development and analysis of multi-layer scaffolds for tissue engineering. Biomaterials 30:6228-6239

114. Peter WD, Vicki S, Richardson JC (2011) The key role alginates play in health. Food Hydrocolloids 25:263-266

115. Philip S, Keshavarz T, Roy I (2007) Polyhydroxyalkanoates: biodegradable polymers with a range of applications. J Chemical Tech Biotech 2(3):233-247

116. Picker KM, Hoag SW (2002) Characterization of the thermal properties of microcrystalline cellulose by modulated temperature differential scanning calorimetry. J Pharmaceutical Sci 91:342-349

117. Prashant RC, Ishwar BB, Shrikant AS, Rekha SS (2009) Microbial cellulose: fermentive production and applications. Food Technol Biotechnol 47:107-124

118. Qin Y, Cai L, Feng D, Shi B, Liu J, Zhang W, Shen Y (2007) Combined use of chitosan and alginate in the treatment of waste water. J Appl Polym Sci 104:3181-3587

119. Rafael A, Loong TL, Susan EM, Selke HT (2010) Poly(lactic acid): synthesis, structures, properties, processing and applications. Chapter 28:457-467

120. Rajev AJ (2000) The manufacturing techniques of various drug loaded biodegradable poly(lactide-co-glycolide) (PLGA) devices. Biomaterials 21:2475-2490

121. Ramya R, Venkatesan , Jayachanndran Kim S, Sudha PN (2012) Biomedical applications of chitosan: an overview. J Biomaterial Tissue Engg 2:100-111

122. Ravenstijn JTJ (2010) The state-of-the art on bioplastics: products, markets, trends and technologies. Polymedia, Lüdenscheid.

123. Ray SS, Bousmina M (2005) Biodegradable polymers and their layered silicate nanocomposites: in greening the 21st century materials world. Progress Material Sci 50:962-1079

124. Reis KC, Pereira J, Smith AC, Carvalho CWP, Wellner N, Yakimets I (2008) Characterization of polyhydroxybutyrate-hydroxyvalerate (PHB-HV)/maize starch blend films. J Food Engg 89:361-369

125. Rekha MR, Chrndra PS (2007) Pullulan as a promising biomaterial for biomedical applications: a perspective. Trends in Biomaterials and Artificial Organs 20:21-45

126. Roberts GAF (1997) Chitosan production routes and their role in determining the structure and properties of the product. In: Domard M, Roberts AF, Vårum KM (eds) Advances in Chitin Science, vol. 2, National Taiwan Ocean University, Taiwan, Jacques Andre, Lyon. pp 22-31 1998

127. Rubin AL, Stenzel KH, Miyata T, White MJ, Dune M (1973) Collagen as a vehicle for drug delivery: preliminary report. J of Clinical Pharmacology 13:309-312

128. Ruth KG, Hartmann R, Egli T, Zinn M, Ren Q (2007) Efficient production of (R)-3-hydroxycarboxylic acids by biotechnological conversion of polyhydroxyalkanoates and their purification. Biomacromolecules 8:279-286

129. Savenkova L, Gercberga Z, Nikolaeva V, Dzene A, Bibers I, Kalina M (2000) Mechanical properties and biodegradation characteristics of PHB-based films. Process Biochem 35:537-579

130. Schmidmaier G, Wildemann A, Stemberger A, Has MR (2001) Biodegradable poly(D, L-lactide) coating of implants for continuous release of growth factors. J Biomed Mater Res 58:449-455

131. Sena RF, Costelli MC, Gibson LH, Coughlin RW (2006) Enhanced production of pullulan by two strains of A. pullulans with different concentrations of soybean oil in sucrose solution in batch fermentations. Brazilian J Chem Eng 2:507-515

132. Siddhesh NP, Edgar KJ (2012) Alginate derivatization: a review chemistry, properties and applications. Biomaterials 33:3279-3305

133. Singh RS, Saini GK, Kennedy JF (2008) Pullulan: microbial sources, production and applications. Carbohydr Polym 73:515-531

134. Sodian R, Hoerstrup SP, Sperling JS, Daebritz S, Martin DP, Moran AM, Kim BS, Schoen FJ, Vacanti JP, Mayer JE (2000) Early in vivo experience with tissue engineered trileaflet heart valves. Circulation 102:22-29

135. Sodian R, Sperling JS, Martin DP (1999) Tissue engineering of a trileaflet heart valve-early in vitro experiences with a combined polymer. Tissue Engg 5:489-494

136. Steinbüchel A, Valentin HE (1995) Diversity of bacterial polyhydroxyalkanoic acids. FEMS Microbiol Lett 128:219-228

137. Tang H, Ishii D, Mahara A, Murakami S, Yamaoka T, Sudesh K, Samian R, Fujita M, Maeda M, Iwata T (2008) Scaffolds from electrospun polyhydroxyalkanoate copolymers: fabrication, characterization, bio absorption and tissue response. Biomaterials 29:1307-1317

138. Teli MD, Chiplunkar V (1986) Role of thickeners in final performance of reactive prints. Textile Dyer Printer 19:13-19

139. Türesin F, Gürsel I, Hasirci V (2001) Biodegradable polyhydroxyalkanoate implants for osteomyelitis therapy: in vitro antibiotic release. J Biomaterials Sci 12(2):195-207 Polymer Edition

140. Valantin MA, Aubron-Olivier C, Ghosn J, Laglenne E, Pauchard M, Schoen H (2003) Polylactic acid implants to correct facial lipoatrophy in HIV-infected patients: results of the open-label study. Vega Aids 17:2471-2477

141. Valappil S, Misra S, Boccaccini A, Roy I (2006) Biomedical applications of polyhydroxyalkanoates, an overview of animal testing an in vivo responses. Expert Rev Med Devices 3:853-868

142. Valerie D, Vinod DV (1998) Pharmaceutical applications of chitosan. Pharmaceutical Sci Technol Today 1:246-253

143. Virginia E, Marie G, Eric P, Luc A (2011) Structure and properties of glycerol plasticized chitosan obtained by mechanical kneading. Carbohydrate Polym 83:947-952

144. Volova T, Shishatskaya E, Sevastianov V, Efremov S, Mogilnaya O (2003) Results of biomedical investigations of PHB and PHB/PHV fibers. Biochem Eng J 16:125-133

145. Vona IA, Costanza JR, Cantor HA, Robert WJ (1965) Manufacture of plastics, vol 1. Wiley, New York. pp 141-142

146. Walle GAM, de Koning GJM, Weusthuis RA, Eggink G (2001) Properties, modifications and applications of biopolyesters. Adv Biochem Eng Biotechnol 71:264-291

147. Wang Z, Itoh Y, Hosaka Y, Kobayashi I, Nakano Y, Maeda I, Umeda F, Yamakawa J, Kawase M, Yagi K (2003) Novel transdermal drug delivery system with polyhydroxyalkanoate and starburst polyamidoamine dendrimer. J Biosci and Bioengg 95(5):541-543

148. Weraporn PA, Sorapong P, Narongchai OC, Ubon I, Puritud J, Sommai PA (2011) Preparation of polymer blends between poly (L-lactic acid), poly (butylene succinate-co-adipate) and poly (butylene adipate-co-terephthalate) for blown film industrial application. Energy Procedia. 9:581-588

149. Westermark S, Juppo AM, Kervinen L, Yliruusi J (1999) Microcrystalline cellulose and its microstructure in pharmaceutical processing. Eur J Pharm Biopharm 48:199-206

150. Williams SF, Martin DP, Horowitz DM, Peoples OP (1999) PHA applications: addressing the price performance issue I. Tissue engineering. Int J Biol Macromolecules 25:111-121

151. Win NN, Stevens WF (2001) Shrimp chitin as substrate for fungal chitin deacetylase. Appl Microbiol Biotechnol 57:334-341

152. Xie ZP, Huang Y, Chen YL, Jia Y (2001) A new gel casting of ceramics by reaction of sodium alginate and calcium iodate at increased temperature. J Mat Sci Lett 20:1255-1257

153. Yan YF, Krishnaiah D, Rajin M, Bono A (2009) Cellulose extraction from palm kernel cake using liquid phase oxidation. J Engg Sci Tech 4:57-68

154. Yang F, Li X, Li G, Zhao N, Zhang X (2002) Study on chitosan and PHBHHx used as nerve regeneration conduit material. J Biomedical Engg 19:25-29

155. Zajic JE, LeDuy A (1973) Flocculant and chemical properties of a polysaccharide from Pullularia pullulans. Appl Microbiol 25:628-635

156. Zeeshan K, Gopiraman M, Yuichi H, Kai W, Kim IS (2013) Cationic-cellulose nanofibers: preparation and dyeability with

were used in the experiments and the results were compared with pure water. The experimental results showed that the gas (air) hold-up and the volumetric gas-liquid oxygen transfer coefficient values for the micro-emulsion systems were usually greater than those of pure water. The packing installation increased the overall gas-liquid volumetric mass transfer coefficient by increasing the flow turbulence and Reynolds number, compared to the unpacked column. The packing increased the gas hold-up and decreased the bubble size and liquid circulation velocity. Furthermore, two empirical correlations were developed to predict the overall gas hold-up and volumetric oxygen transfer coefficient. A good agreement was observed between the experimental and correlated data.

INTRODUCTION

Airlift reactors are a modified type of bubble column in which the internal structure is divided into two separate sections by a baffle split or draft tube, which are the riser and down-comer (Carvalho, 2000et al., 2001; Hekmat et al., 2010).

The gas hold-up (ε) difference between the riser and down-comer of an airlift reactor creates a density difference between these two zones as a driving force for liquid circulation (Moraveji et al., 2012a).

Petroleum refineries generate large amounts of wastewater with high concentrations of hazardous contaminants. The effluents of these refineries usually contain 1-10% oil, the bulk consisting of water, emulsifiers and other ingredients (Jonsson and Tragardh, 1990). Many industries such as petrochemical, petroleum production, oil refinery factories, metal, food industries and cosmetics usually produce a wastewater containing emulsions of oil in water (O/W) (Kong and Li, 1999; Lin and Lan, 1998).

Because of the toxic nature and important effects of different kinds of oily wastewater on the surrounding environment (soil, water), it is necessary to treat the wastewaters before discharging to the environment. There are several methods such as dissolved air flotation (DAF) (Al-Shamrani et al., 2002; Bensadok et al., 2007), adsorption (Ayotamuno et al., 2006), biological treatment (Tellez et al., 2002; Zhao et al., 2006), and sedimentation in a centrifugal field (Cambiella et al., 2006) and in hydro-cyclones (Hashmi et al., 2004) that can be used for the treatment of wastewater. Since the upper limit of permissible oil in the effluent is steadily reduced by governing bodies, industries have to improve the treatment plants to meet the new limits.

The oil-in-water emulsion is a colloidal dispersion of oil droplets in an aqueous medium. These droplets have a tendency to coalesce and separate from the aqueous phase.

One of the useful equipments that can treat wastewaters is the airlift reactor. Most of the published works on airlift reactors have been on air-water based systems with properties different from the real conditions of the aerobic bio-desulfurization (BDS) processes. There are some reports on BDS in airlift reactors (Monticello, 2000; Lange and Pacheco, 1999) which mostly are water-in-oil (W/O) micro-emulsions.

Mehrnia et al. (2004a & 2004b) used a water-inkerosene (W/K) micro-emulsion at a water-to-oil phase volume ratio (φ) of 20%, as a cold model of the BDS inside a draft-tube airlift bioreactor (DTAB) with different geometries. They also correlated their results using viscosity and aeration velocity (UG), although their variables were reported for the homogenous flow regime.

In another research, the viscosity effect on the overall volumetric oxygen transfer coefficient (kLa) using water-in-diesel (W/D) micro-emulsions with a water to diesel volume ratio (φ) in the range 2.5–40% inside the same reactor was investigated (Shariati et al., 2007). They found that an increase of the viscosity of the micro-emulsion resulted in a further reduction of the values of kLa at all air rates for which measurements were done.

However, several techniques were applied to enhance the oxygen transfer rates in columns at low aeration rates (Fraser and Hill, 1993; Godo et al. 1999; Su and Heindel, 2004), but the oxygen transfer coefficient was significantly increased by adding packing into the column (Moraveji et al., 2011a).

Nikakhtari and Hill (2005) used stainless steel meshes as a packing with 99.0% porosity in the riser section of an external loop airlift bioreactor (ELAB). According to this research, the overall volumetric oxygen transfer coefficient increased (by an average factor of 2.45) in a packed bed ELAB compared to the same ELAB without packing. The overall volumetric oxygen transfer coefficient reached 0.021 s^{-1} at a gas superficial velocity of 0.0157 m/s in a packed reactor.

The aim of this research is to investigate the effect of aeration velocity and liquid properties on the hydrodynamic parameters and volumetric mass transfer coefficient in a packed split-cylinder airlift reactor. The packing was installed in the riser section of the airlift reactor.

Four different oil-in-water micro-emulsions containing kerosene, heavy naphtha, light naphtha, and diesel as the oil at a concentration of 7 % (v/v) were prepared and their behavior was carefully studied. The surface tension of the bulk liquid decreases and smaller bubbles were produced in microemulsions in comparison with pure water. Therefore, the gas hold-up and mass transfer coefficient increased in micro-emulsions; however, adversely liquid circulation velocity decreased.

EXPERIMENTAL

Materials and Methods

Different petroleum fractions containing kerosene, heavy naphtha, light naphtha and diesel were purchased from Shazand Oil refinery Company (Arak, Iran) and their various solutions at a concentration of 7% (v/v) were prepared locally. Micro-emulsions were prepared from tap water and petroleum fractions. Nonyil Phenol [(NF-60) purity: 99.5% purchased from Isfahan Copolymer Company] was added as the emulsifier. The emulsifier, petroleum fractions and liquid properties are respectively summarized in Tables 1-3.

Table 1: Physical properties of the emulsifier NF-60

HLB	pH	Cloud point (°C)	Density at 20 °C (g/ cm³)	Water wt%	Avg M.W.	Appearance	Avg EO mole	Trade name
10.9	5-7	60±4	1.045±0.01	0.5 Max	484	Oily liquid	6	NF-60

Table 2: Physical properties of the oil fractions at 20 °C

RSH (ppm)	H$_2$S	Pour point	Flash point	FBP (°C)	IBP	Sp gravity 15.5/15.5 °C	Trade name
<10	FREE	—	—	89	47	0.668	L. Naphtha
<10	FREE	—	—	157	95	0.7495	H. Naphtha
<10	FREE	—	49	257	161	0.8035	Kerosene
<10	FREE	3	111	380	239	0.8265	Diesel

Table 3: Physical properties of liquid

Liquid	Oil-in-water volume	Density (kg/m^3	Kinematic viscosity 10^{-6} (m^2/s)	Surface tension (mN/m)
Water	—	998.2	0.902	72.8
Water/L. Naphtha	7%	973.829	1.42	24.293
Water/H. Naphtha	7%	979.517	2.11	25.116
Water/ Kerosene	7%	983.286	3.70	26.045
Water/ Diesel	7%	984.891	27.19	27.532

Apparatus Set Up and Measurement Methods

The split-cylinder airlift reactor used in this research was the same as the apparatus applied by Moraveji et al. (2011b), as shown in Figure 1. The main difference between our case and the literature is ceramic Pall Ring packing (as shown in Figure 2), which was inserted in the riser section of the airlift reactor. The packing density, dry factor and specific surface area were 540 kg/m^3, 356 m^{-1} and 150 m^2/m^3, respectively. The maximum possible porosity of the packing is equal to 78.0%. As is known, packing normally increases the mass transfer and gas hold-up. Therefore, packing was applied in this work to compare the unpacked results (Moraveji *et al.*, 2011a) with the packed ones. All experiments were carried out in ambient conditions [atmospheric pressure and 25(± 0.5) °C)].A dissolved oxygen electrode (WTWCellox325) for oxygen concentration measurement in the liquid bulk was set in the riser zone at a depth of 0.1 m from the surface of the gas free liquid. The probe's tip was at an angle of 30° to the horizon to prevent oxygen bubbles from sticking to it. The conductivity electrode (Model 740i, WTW, Germany) for liquid circulation velocity and mixing time measurements was positioned in the down-comer zone at a depth of 0.2 m from the bottom of the reactor.

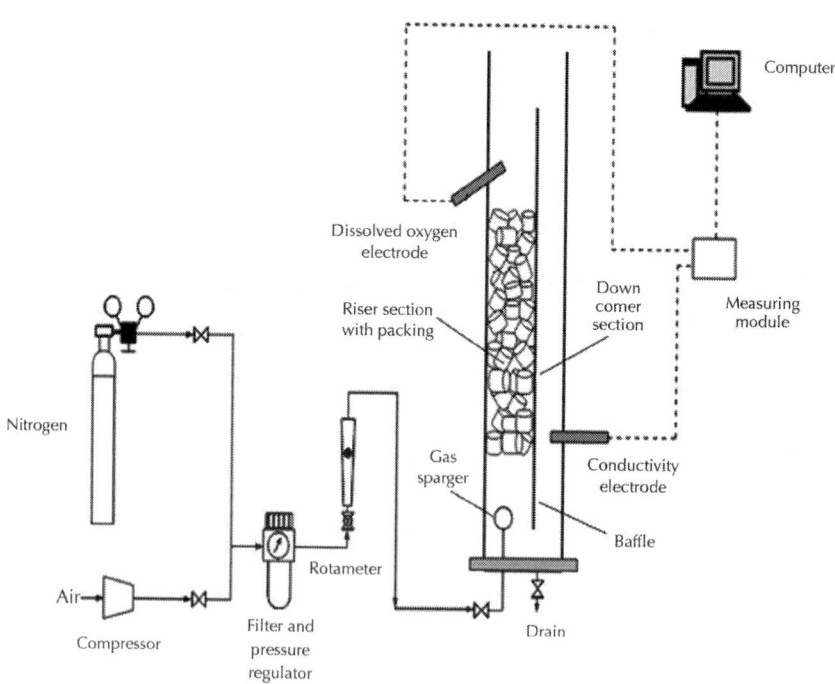

Figure 1: Schematic diagram of the split-cylinder airlift reactor.

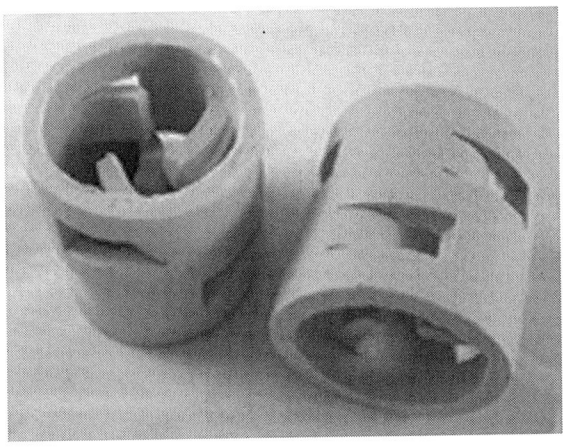

Figure 2: Ceramic Pall Ring packing used in this research.

In this study, the volume expansion method was applied to measure the overall gas hold-up (ε) during the steady state condition in the airlift bioreactor. Liquid circulation velocity (UL) and oxygen mass transfer coefficient measurement methods have been proposed and reviewed elsewhere (Moraveji et al.,2011b).

RESULTS AND DISCUSSION

Overall Gas Hold-Up

Figure 3 shows the overall gas hold-up for four different micro-emulsions versus superficial gas velocity. Three independent parameters involving surface tension (σ), density (ρ) and viscosity (v) affect the micro-emulsions properties in an unpacked airlift reactor (Moraveji et al., 2012b).

According to the literature, the order of the surface tension, density and viscosity for the four different micro-emulsions (at the same concentration) is the following (Moraveji et al., 2012c):

light naphtha < heavy naphtha < kerosene < diesel

As shown in Figure 3, the gas hold-up originally increased with an increase of the superficial gas velocity. Furthermore, the gas hold-up for oil-in water micro-emulsion systems is significantly higher than that of the pure water system. This is due to the reduction of surface tension and the increased hindering of coalescence characteristic of the micro-emulsion systems (compared to pure water).

Therefore, in the micro-emulsion systems small bubbles are created and gas hold-up increases.

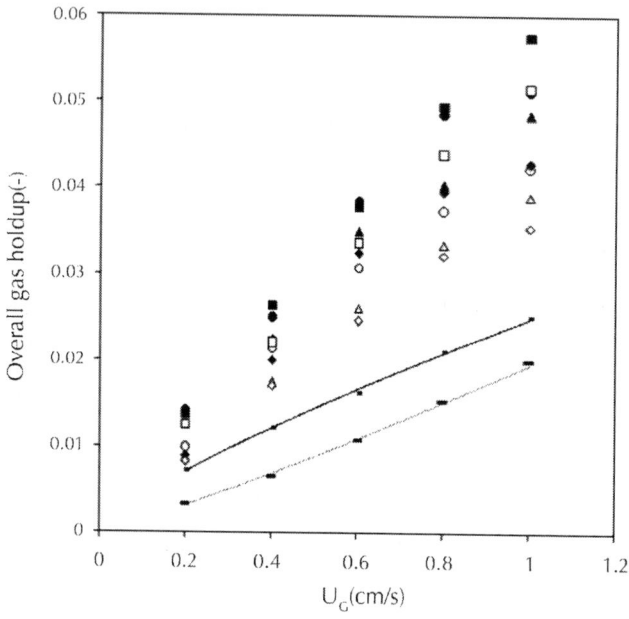

Figure 3: Overall gas hold-up versus superficial air velocity (UG).

(◆ Diesel 7%; ■ Light Naphtha 7%; ▲ Kerosene 7%; . Heavy Naphtha 7%; - pure water) Packed. (◇ Diesel 7%; Light Naphtha 7%; △ Kerosene 7%; Heavy Naphtha 7%; - pure water) Unpacked.

In the solutions with higher densities, buoyancy force enhancement helps the bubbles to rise, so the gas hold-up decreases. Although viscosity enhancement slightly increases the bubble residence time and gas hold-up, it creates bigger bubbles and increases the buoyancy force. Therefore, the gas hold-up decreases.

Further, packing decreases the gas bubble size. Some bubbles are captured inside the packing, so the gas hold-up increases.

According to our research, the maximum gas hold-up in a packed bed reactor was obtained for a light naphtha-in-water micro-emulsion. It was 60%more than that of pure water at the highest aeration velocity (UG=1cm/s). Therefore, the overall gas hold-up increased as follows:

water < diesel < kerosene < heavy naphtha < light naphtha

According to our conditions in this research, the gas hold-up was correlated by applying dimension less numbers as:

$$\frac{\varepsilon}{(1-\varepsilon)^4} = 0.42Bo^{0.26}Ga^{0.08}Fr$$

(1)

where, Bo ,Ga and Fr are the Bond number [the ratio of body forces (which often is equal to the gravity forces) to the surface tension forces], the Galilei number (the ratio of gravity forces to viscous forces) and the Froude number (the ratio of inertial forces to the gravity forces), respectively.

A comparison between the experimental data and the correlated results for the gas hold-up is illustrated in Figure 4. As shown in this figure, Eq. (1) can predict the data with 16% error. Sada et al.'s equation showed a good agreement with the experimental data while Akita and Yoshida's equation showed some errors. This is due to the Bo number effect.

Table 4 compares two correlations obtained from the literature (Sada et al., 1984; Akita and Yoshida, 1974) and our correlation.

Liquid Circulation Velocity

It was experimentally found that the liquid circulation velocity (UL) strongly depends on the gas hold-up. The difference of gas hold-up between the riser and the down-comer zones of the airlift reactor provides the driving force for the liquid circulation.

Figure 5 shows the experimental data for the velocity of bulk liquid circulation in the bioreactor versus the superficial gas velocity in the riser for water and various micro-emulsions employed in this study. The gas hold-up sharply increases in the down-comer with a decrease of surface tension and bubble diameter. Therefore, the difference of gas hold-up between the riser and down-comer decreases. On the other hand, the driving force for the liquid circulation decreases. So, the velocity of the micro-emulsions liquid circulation velocity is less than that of pure water. The liquid circulation velocity increases as:

light naphtha < heavy naphtha < kerosene < diesel < water

At the highest aeration velocity (UG=1cm/s), the minimum liquid circulation velocity was observed for light naphtha (30% less than that of pure water) and the maximum liquid circulation velocity was observed for diesel (8.5% less than that of pure water).

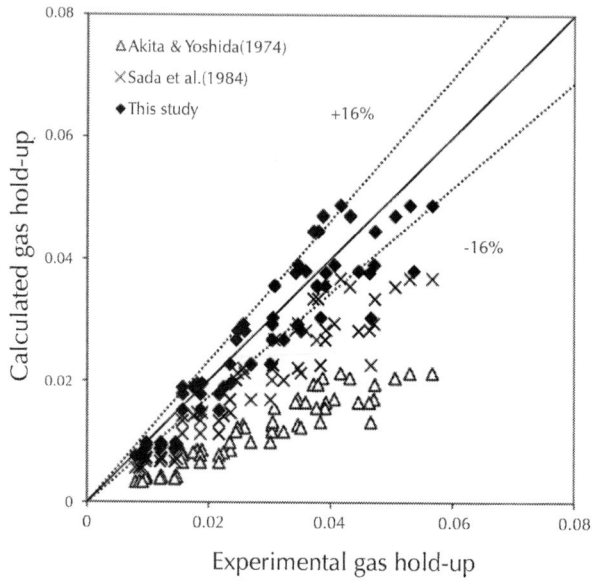

Figure 4: The correlated data versus experimental gas hold-up.

Table 4: Gas hold-up correlations based on Bo number, Ga number and Fr number.

Correlations	Remarks	References
$\dfrac{\varepsilon}{(1-\varepsilon)^4} = 0.2Bo^{1/8}Ga^{1/12}Fr$	$d_b > 2.5mm$	Akita &Yoshida (1974)
$\dfrac{\varepsilon}{(1-\varepsilon)^4} = 0.32Bo^{0.21}Ga^{0.086}Fr$	$0.3 < U_G < 30cm/s$	Sada et al. (1984)
$\dfrac{\varepsilon}{(1-\varepsilon)^4} = 0.42Bo^{0.26}Ga^{0.08}Fr$	$0.2 < U_G < 1cm/s$	This study

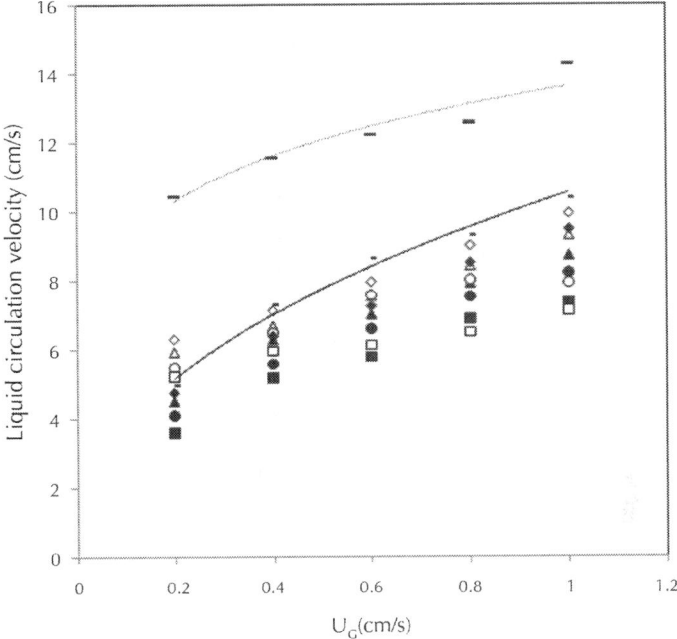

Figure 5: Liquid circulation velocity in the riser–down-comer loop versus aeration velocity (UG). (◆ Diesel 7%; ■ Light Naphtha 7%; ▲ Kerosene 7%; Heavy Naphtha 7%; - pure water) Packed. (◇ Diesel 7%; Light Naphtha 7%; △ Kerosene 7%; Heavy Naphtha 7%; - pure water) Unpacked.

Mixing Time

Figure 6 shows the mixing time (tm) versus superficial gas velocity for micro-emulsions and pure water. The mixing time decreased by increasing the aeration velocity in the riser. As shown in this figure, the mixing time increased as follows:

water < diesel < kerosene < heavy naphtha < light naphtha

The mixing time enhancement is due to the decrease of surface tension and bubble diameter in micro-emulsions. Circulation of bubbles in the reactor assists the mixing process. Packing installation has a negative impact on the magnitude of the induced liquid circulation velocity and increases the mixing time.

Mass Transfer

Figure 7 shows the volumetric mass transfer coefficient (kLa) for pure water and four different micro-emulsions versus the superficial gas velocity. As shown in this figure, kLa values increased upon increasing the superficial gas velocity. The kLa values for all of the micro-emulsions were significantly greater than that of pure water, as:

water < diesel < kerosene < heavy naphtha < light naphtha

However, the maximum kLa was obtained for the light naphtha micro-emulsion (about 31% more than that of pure water), but the minimum kLa was obtained for the diesel micro-emulsion (about 4.7% more than that of pure water) (at the highest superficial air velocity).

The surface tension of water is higher (more than three times) than that of the oily micro-emulsions. This causes a bubble diameter reduction of micro-emulsions. The gas-liquid interfacial area (a) increases due to bubble diameter reduction. Therefore, the mass transfer increases.

Further, a liquid viscosity enhancement increases the bubble coalescence rate. It means that the interfacial area decreases in size. The liquid viscosity enhancement thickens the liquid boundary layer bubbles (Appasani, 2007) and decreases the solute diffusivity (Wilke and Chang, 1955). Therefore, kLavalues decrease upon increasing the viscosity of the oily micro-emulsions (Kilonzo and Margaritis, 2004; Calderbank, 1967).

Density is another bulk liquid property that plays a major role in the overall mass transfer. By increasing density, the buoyancy force increases. Therefore, bubbles can rise easily and rapidly. Furthermore, the necessary time for mass transfer between the two phases decreases upon increasing the density. In addition, the packing increases the gas hold-up, and decrease the liquid circulation velocity and gas bubble size in an airlift reactor.

The volumetric mass transfer coefficient for oil in-water micro-emulsions was correlated using theSh number with the following equation:

$$Sh = 0.14 Re^{2/3} Sc^{1/2} Bo^{0.98} \tag{2}$$

where Sh, Re and Sc are the Sherwood number (the ratio of convective and diffusive mass transfer coefficients), the Reynolds number (the ratio of inertial forces to viscous forces) and the Schmidt number (the ratio of momentum diffusivity(viscosity) to mass diffusivity), respectively. For this equation, the coefficient of determination (R2) was around 0.96.

However, our correlation was in good agreement with the results obtained by Asgharpour *et al.* (2010) and Akita and Yoshida (1974), but the Bird *et al.* (2002) correlation was not in good agreement with our experimental data. This may be due to the Bond number influence. In such solutions, the liquid density, surface tension and bubble diameter clearly change, so the Bo number affects the results and cannot be neglected. All of the correlations are illustrated in Figure 8 and compared with our correlated data. Table 5 also shows the corrected correlations based on the literature (Akita and Yoshida, 1974; Asgharpour *et al.*, 2010; Bird *et al.*, 2002) and our correlation.

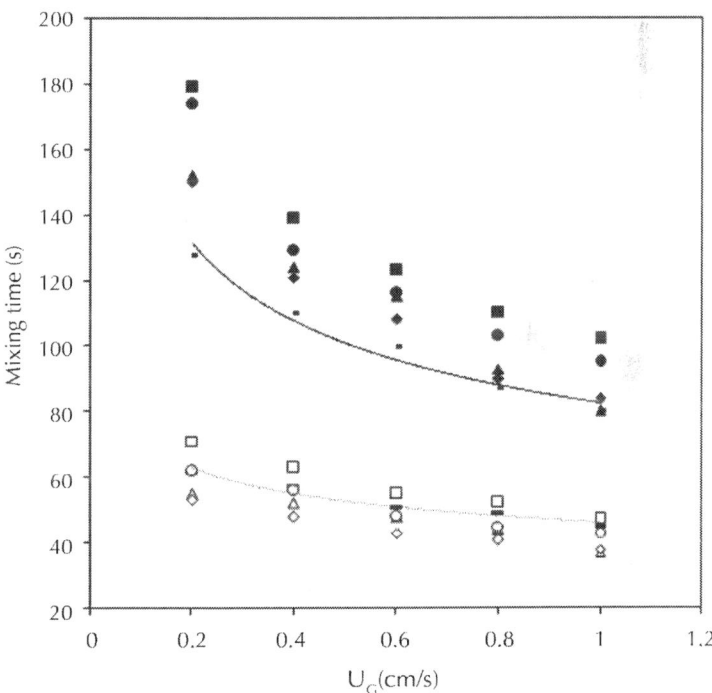

Figure 6: Mixing time versus superficial air velocity UG in the riser.

(◆ Diesel 7%; ■ Light Naphtha 7%; ▲ Kerosene 7%; Heavy Naphtha 7%; - pure water) Packed. (◇ Diesel 7%; Light Naphtha7%; △ Kerosene 7%;

Heavy Naphtha 7%; - pure water) Unpacked.

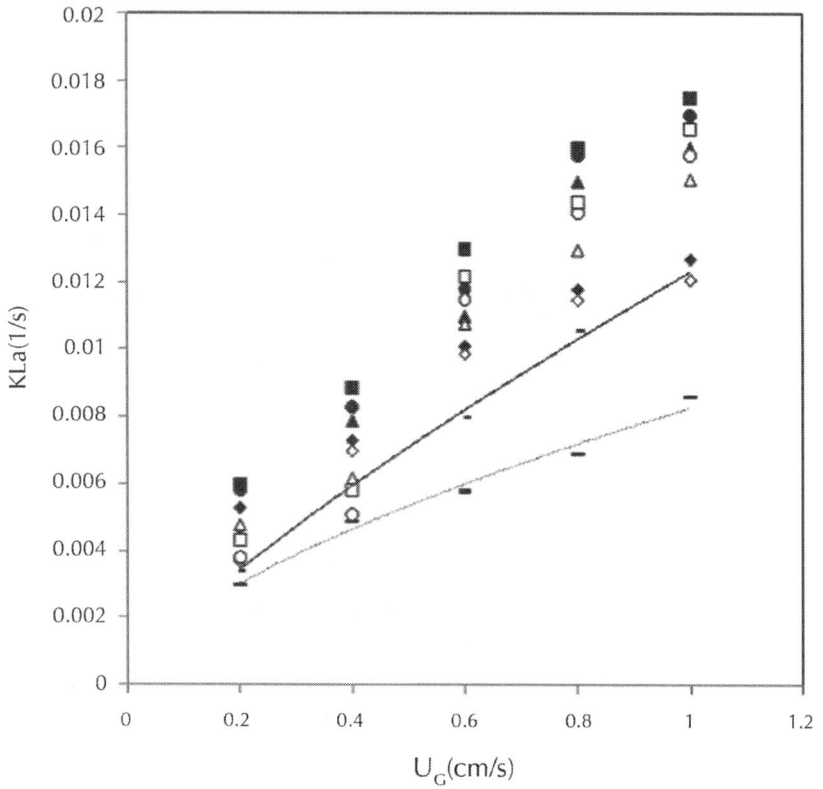

Figure 7: Overall volumetric oxygen mass transfer coefficient (kLa) versus superficial air velocity (UG) in the riser. (◆ Diesel 7%; ■ Light Naphtha 7%; ▲ Kerosene 7%; Heavy Naphtha 7%; - pure water) Packed. (◇ Diesel 7%; Light Naphtha7%; △ Kerosene 7%;

Heavy Naphtha 7%; -: pure water) Unpacked.

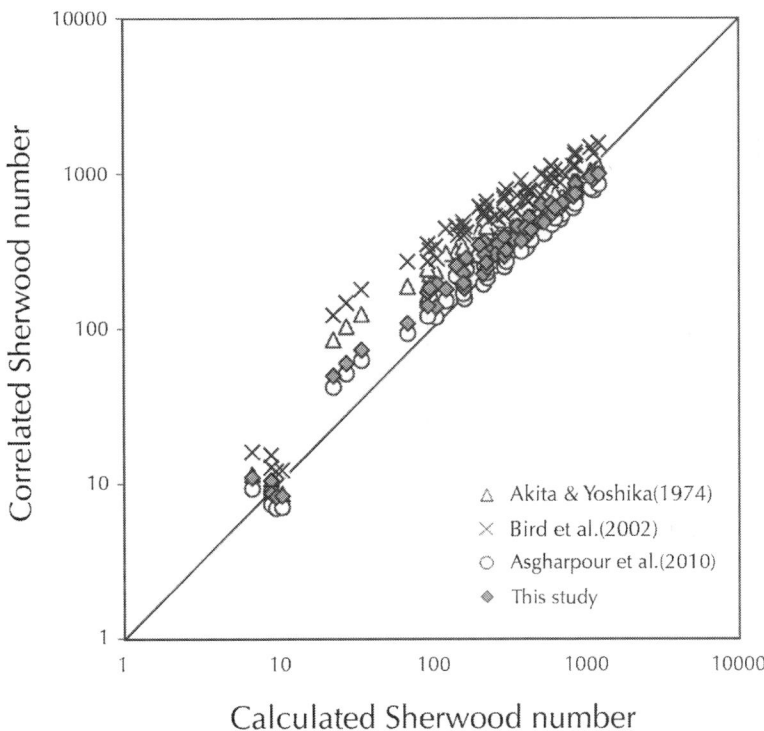

Figure 8: The correlated Sherwood number versus the Sherwood numbers obtained from the experimental work.

Table 5: Sherwood number correlations for the liquid-gas mass transfer coefficient based on Re number, Sc number and Bo number

Correlations	Remarks	References
$Sh = 0.15 Re^{2/3} Sc^{1/2} Bo^{2/3}$	$0.118 < U_G < 2.35$ cm/s	Asgharpour et al. (2010)
$Sh = \left(\dfrac{4}{\pi}\right)^{1/2} Re^{1/2} Sc^{1/2}$	Higbie's model	Bird et al. (2002)
$Sh = 0.6 Re^{1/2} Sc^{1/2} Bo^{3/8}$	Homogeneous flow	Akita & Yoshida (1974)
$Sh = 0.14 Re^{2/3} Sc^{1/2} Bo^{0.98}$	$0.2 < U_G < 1$ cm/s	This study

CONCLUSIONS

In this study, gas hold-up, liquid circulation velocity, mixing time and the volumetric oxygen transfer coefficient were considered in a packed split-cylinder airlift reactor and compared with the unpacked reactor data. Four different oil-in-water micro-emulsions [containing light naphtha, heavy naphtha, kerosene and diesel at a concentration of7% (v/v)] were used and compared with the pure water system. The light naphtha based micro-emulsion with the lowest surface tension and viscosity had the maximum mass transfer coefficient and gas hold-up and had the minimum liquid circulation velocity, while diesel with the highest surface tension and viscosity had the minimum mass transfer coefficient and gas hold-up and had the maximum liquid circulation velocity. Since gas bubbles were captured inside the packing, the mass transfer increased compared to an unpacked reactor. Furthermore, these bubbles increased the gas holdup. The bubble size was more uniform and they had smaller diameters after passing through the packing. In addition, for the micro-emulsion systems in an airlift reactor, two appropriate correlations based on dimensionless groups (Bo, Ga and Fr for gas hold-up and Bo, Re and Sc for Sh) were developed. A satisfactory agreement between the correlated data and experimental results was observed.

REFERENCES

1.   Akita, K., Yoshida, F., Bubble size, interfacial area and liquid-phase mass transfer coefficient in bubble columns. Ind. Eng. Chem. Proc. Des. Dev., 13, 84-91 (1974).

2.   Al-Shamrani, A. A., James, A., Xiao, H., Separation of oil from water by dissolved air flotation. Colloids Surf., 209, 15-26 (2002).

3.   Appasani, P., CX CFD simulation of the hydrodynamics and inter-phase mass transfer in airlift reactors. Master Thesis, Dalhousie University (2002).

4.   Asgharpour, M., Mehrnia, M. R., Mostoufi, N., Effect of surface contaminants on oxygen transfer in bubble column reactors. Biochem. Eng. J., 49,351-360 (2010).

5. Ayotamuno, M. J., Kogbara, R. B., Ogaji, S. O. T., Pobert, S. D., Petroleum contaminated groundwater: Remediation using activated carbon. Appl. Energy, 83, 1258-1264 (2006).

6. Bensadok, K., Belkacem, M., Nazzal, G., Treatment of cutting oil/water emulsion by coupling coagulation and dissolved air flotation. Desalination, 206, 440-448 (2007).

7. Bird, R. B., Stewart, W. E., Lightfoot, E. N., Transport Phenomena. 2nd Ed. John Wiley and Sons Inc. New York, USA (2002).

8. Calderbank, P. H., Mass Transfer in Fermentation Equipment. In: N. Blakeborough (Ed.). Biochemical and Biological Engineering Science, Vol. 1,Academic Press, London (1967).

9. Cambiella, A., Bentio, J. M., Pazos, C., Coca, J., Centrifugal separation efficiency in the treatment of waste emulsified oils. Trans. IChemE, 84, 69-76 (2006).

10. Carvalho, E., Camarasa, E., Meleiro, L. A. C.,Maciel Filho, R., Domingues, A., Vial, Ch., Wild,G., Poncin, S., Midoux, N., Bouillard, J., Development of a hydrodynamic model for airlift reactors. Braz. J. Chem. Eng., 17, 607-616(2000).

11. Fraser, R. D. and Hill, G. A., Hydrodynamic characteristics of a spinning sparger, external loop airlift bioreactor. Can. J. Chem. Eng., 71, 419-425(1993).

12. Godo, S. Klein, J. Polakovic, M. and Bales, V., Periodical changes of input air flow rate, a possible way for improvement of oxygen transfer and liquid circulation in airlift bioreactors. Chem. Eng. Science, 54, 4937-4943 (1999).

13. Hashmi, K. A., Hamza, H. A., Wilson, J. C.,CANMET hydrocyclone: An emerging alternative for the treatment of oily waste streams. Minerals Eng. 17, 643-649 (2004).

14. Hekmat, A., Ebadi, Amooghin, A., KeshavarzMoraveji, M., CFD simulation of gas–liquid flow behaviour in an air-lift reactor: Determination of the optimum distance of the draft tube. Sim. Model. Pract. Theo., 18, 927-945 (2010).

15. Jonsson, A. S., Tragardh, G., Ultrafiltration applications. Desalination, 77, 135-179 (1990).

16. Kilonzo, P. M., Margaritis, A., The effects of non-Newtonian fermentation broth viscosity and small bubble segregation on

oxygen mass transfer in gas-lift bioreactors: A critical review. Biochem. Eng. J., 17, 27-40 (2004).

17. Kong, J., Li, K., Oil removal from oil-in-water emulsions using PVDF membranes. Sep. Purif. Technol., 16, 83-93 (1999).

18. Lange, E. A., Pacheco, M. A., Advances in biocatalytic desulphurization. Petroleum Technol. Quarterly (Autumn), 37-43 (1999).

19. Lin, S. H., Lan, W. J., Waste oil/water emulsion treatment by membrane processes. J. Hazard. Materials, 59, 189-199 (1998).

20. Mehrnia, M. R., Towfighi, J., Bonakdarpour, B.,Akbarnejad, M. M., Design and operational aspects of airlift bioreactors for petroleum biodesulfurization. Environmental Prog., & Sustainable Energy,23, 206-214 (2004a).

21. Mehrnia, M. R., Towfighi, J., Bonakdarpour, B.,Akbarnejad, M. M., Influence of top-section design and draft-tube height on the performance of airlift bioreactors containing water-in-oil micro-emulsion. J. Chem. Technol. and Biotechnol., 79, 260-267 (2004b).

22. Monticello, D. J., Biodesulfurization and the upgrading of petroleum distillates. Curr. Opin. in Biotechnol.,11, 540-546 (2000).

23. Moraveji, M. K., Sajjadi, B., Jafarkhani, M., Davarnejad, R., Experimental investigation and CFD simulation of turbulence effect on hydrodynamic and mass transfer in a packed bed airlift internal loop reactor. Int. Commun. Heat and Mass Trans., 38, 518-524 (2011a).

24. Moraveji, M. K., Sajjadi, B., Davarnejad, R., Gas liquid hydrodynamics and mass transfer in aqueous alcohol solutions in a split-cylinder airlift reactor. Chem. Eng. Technol., 34, 465-474 (2011b).

25. Moraveji, M. K., Morovati Pasand, M., Davarnejad, R., Chisti, Y., Effects of surfactants on hydrodynamics and mass transfer in a split-cylinder airlift reactor. Can. J. Chem. Eng., 90, 93-99 (2012a).

26. Moraveji, M. K., Mohsenzadeh, E., EbrahimiFakhari, M., Davarnejad, R., Effects of surface active agents on hydrodynamics and mass transfer characteristics in a split-cylinder airlift

bioreactor with packed bed. Chem. Eng. Res. Des., 90, 899-905 (2012b).

27. Moraveji, M. K., Jafarkhani, M., Sajjadi, B.,Davarnejad, R., Hydrodynamics and mass transfer of oil–water micro-emulsion in a three phase internal airlift reactor. Fuel, 97, 197-201 (2012c).

28. Nikakhtari, H., Hil, G. A., Hydrodynamic and oxygen mass transfer in an external loop airlift bioreactor with a packed bed. Biochem. Eng. J., 27,138-145(2005).

29. Sada, E., Katoh, S., Yoshil, H., Performance of the gas-liquid bubble column in molten salt system.Ind. Eng. Chem. Proc. Dev., 23, 151-154 (1984).

30. Shariati, F. P., Bonakdarpour, B., Mehrnia, M. R.,Hydrodynamics and oxygen transfer behavior of water in diesel micro-emulsions in a draft tube airlift bioreactor. Chem. Eng. and Proc., 46, 334-342 (2007).

31. Su, X. and Heindel, T. J., Gas hold-up behavior in nylon fiber suspensions. Ind. & Eng. Chemistry Res., 43, 2256-2263 (2004).

32. Tellez, G. T., Nirmalakhandan, N., Gardea-Torresdey, J. L., Performance evaluation of activated sludge system for removing petroleum hydrocarbons from oilfield produced water. Adv. in Environmental Res., 6, 455-470 (2002).

33. Wilke, C. R., Chang, P., Correlation of diffusion coefficients in dilute solutions. AIChE J., 1, 264-270 (1955).

34. Zhao, X., Wang, Y., Ye, Z., Borthwick, A. G. L., Ni,J., Oil field wastewater treatment in biological aerated filter by immobilized microorganisms. Proc. Biochemistry, 41, 1475-1483 (2006).

Sediment Removal from Crude Oil Emulsion Using Microwave Radiation

Fabiane G. AntesI; Juliana S. F. PereiraI; Liange O. DiehlI; Letícia S. F. PereiraII; Paula BoeckI; Regina C. L. GuimarãesIII; Ricardo A. GuarnieriIII; Bianca M. S. FerreiraIII; Maria de Fatima P. dos SantosIV; Edson Luiz FolettoII; and Erico M. M. FloresI

IDepartamento de Química, Universidade Federal de Santa Maria, 97105-900 Santa Maria-RS, Brazil

IIDepartamento de Engenharia Química, Universidade Federal de Santa Maria, 97105-900 Santa Maria-RS, Brazil

IIICentro de Pesquisas e Desenvolvimento Leopoldo Américo Miguez de Mello, Petrobras/CENPES, 21941-945 Rio de Janeiro-RJ, Brazil

IVDepartamento de Ciências Matemática e Naturais, Universidade Federal do Espírito Santo, 29932-540 São Mateus-ES, Brazil

ABSTRACT

Microwave radiation in closed vessels was applied for removal of sediments from crude oil allowing the subsequent crude oil characterization in laboratory. Heating time and microwave power were evaluated in the range of 5 to 30 s and 300 to 1400 W, respectively. Sediment content was determined using the method recommended by ASTM D 4807-05 based on a filtration step in a membrane before and after sediment removal. Water and chloride contents were determined in the oil phase obtained after sediment removal. Up to eight samples of 20 g of crude oil could be simultaneously processed. Sediment removal efficiency was better than 95%. It was possible to obtain crude oil in a suitable condition for determination of routine parameters as API gravity, density, viscosity and total acid number without interferences caused by sediments or even water and salt.

INTRODUCTION

Separation of water from crude oil is a challenging task in oilfields due to the high stability of water-in-oil (W/O) emulsions.[1] It is already known that emulsion stability is directly related to oil composition and naturally occurring compounds present in crude oil.[2] Asphaltenes and resins have been shown to promote and stabilize crude oil emulsion. Additionally, some works have demonstrated that inorganic solids can increase the emulsion stability. The influence of these emulsion stabilizers in crude oil was proved using model and natural emulsions and it was concluded that solids play one of the major effects in the stabilization of W/O emulsions.[3,4]

The mechanism of solids in the stabilization of W/O emulsions is contradictory. According to Sullivan et al.,[2] the presence of inorganic solids could increase the emulsion stability if the particle size is small enough to become interfacially active to cause the adsorption of resins and asphaltenes. The decrease of particle size would enhance the emulsion stability due to increase of interactions between asphaltenes and particles. On the other hand, it was concluded that solids have no effect in the adsorption of asphaltenes on water droplet surface, in spite of their contribution for emulsion stability.[1]

Many efforts have been performed to develop new strategies for achieving an efficient breaking of W/O emulsions in crude oil and, consequently, to remove water, salt and solids. However, despite the importance of demulsification process, the laboratory characterization of crude oil can be more challenging than the step of the emulsion separation.[5] The characterization of crude oil is extremely important for decisions during exploration, production and refining steps. In this way, samples with low content of water and salt are normally required for the determination of crude oil characteristics, such as density, viscosity, among others.[6,7] In this sense, it is necessary to remove water, salts and also solids from crude oil emulsions before laboratory characterization in order to avoid interferences during the analysis.

Water removal in laboratory scale is currently performed using conventional heating and chemical additives since the efficiency of removal can be evaluated by measuring the water content in the oil phase or the salt content in the aqueous phase.[8] Other techniques have been investigated for water removal such as settling by gravity and simple distillation.[9,10] In spite of relatively good efficiency of these methods, they are time consuming, especially for heavy crude oils, and the addition of demulsifiers is frequently required.[11] Additionally, salt and solids are not efficiently removed from the crude oil and can cause interferences especially on density (API gravity) determination.[7]

The determination of sediments in crude oil is usually performed using a method recommended by the American Society for Testing and Materials (ASTM D 4807-05), that is based on a dilution step of samples with toluene, filtration through a 0.45 μm membrane and quantification of solids by weighing.[12] The solids present in crude oil are commonly described as clay particles.[13] However, recent studies have been done in view of solid particles characterization isolated from crude oil samples using the proposed ASTM method and some authors found that these particles were composed mostly of sodium chloride, called salt spheres.[5,14] The characterization of solid particles also revealed other finely divided materials that were composed of calcium and sulfur atoms, indicating the presence of calcium sulfate.[5,14]

In addition to the determination of sediment and water contents, another important parameter to be determined in the crude oil is the salt content. In this sense, a method that consists of extraction using a mixture of solvent and subsequent chloride determination by titration with AgNO$_3$ is usually employed for the determination of salt (expressed as NaCl) in crude oil.[15] However, although some improvements have been proposed, this method is sample destructive and relatively time consuming.[16] The use of microwave radiation in demulsification processes has been reported in several works.[16,17] Microwave electromagnetic field causes fast heating due to molecular rotation and movement of ions as a consequence of induced or permanent molecule dipoles. Thus, the application of microwave radiation to emulsions results in destabilization of interfacial film because of the temperature increase and consequently water droplet coalescence.[16]

Recently, a method using microwave radiation for salt extraction from heavy crude oils has been proposed.[8] The method involves the addition of water to crude oil in a quartz vessel and subsequent microwave irradiation. Using this method, water and oil phases were completely separated and the salt content was determined in the water phase. On the other hand, the separation of water from heavy crude oil using microwave radiation was subsequently investigated and it was possible to obtain crude oil free of water and salt.[18] The main advantage of the proposed microwave-assisted method is that it does not require chemical demulsifiers or other reagents and enables the determination of the salt content in water phase. Moreover, the demulsified crude oil is obtained in a suitable condition for laboratory characterization. In spite of its good performance for salt extraction and water separation, the feasibility of this method was not evaluated up to now for sediment removal.

In the present work, microwave radiation was applied for removal of sediments from real crude oil emulsions with relatively high sediment content (1 to 6%). The efficiency of the proposed method was evaluated based on the determination of sediments in heavy crude oil emulsion without treatment and in the oil phase obtained after microwave heating procedure. Additionally, water and Cl were also determined in the oil phase after sediment removal. Microwave heating program, sample mass and number of extraction steps were evaluated. Samples obtained after microwave irradiation procedure were characterized by performing the determination of sediments, water, S, N and Cl (reported as NaCl). In addition, total acid number (TAN), kinematic and dynamic viscosity and API gravity were evaluated in demulsified heavy crude oil only after the proposed microwave-assisted sediment removal method.

EXPERIMENTAL

Instrumentation

A microwave sample preparation system (Multiwave 3000, Anton Paar, Graz, Austria) equipped with eight high-pressure quartz vessels (80 mL) was used for the proposed microwave-assisted sediment removal method. Temperature and pressure were controlled in real time during all experiments. In order to keep a safe operation, the maximum operational temperature and pressure selected were 260 °C and 70 bar, respectively. This equipment was also used for crude oil digestion by microwave-induced combustion (MIC) for further Cl determination (before and after sediment removal).[19,20]

An ion chromatographic system (Metrohm, Herisau, Switzerland) equipped with a pump (IC liquid handling unit), a conductivity detector (model 819), auto sampler (model 813), anion self-regeneration suppressor (model 833) and a sampling loop of 100 µL were used for Cl determination in crude oil before and after microwave-assisted sediment removals and after microwave-induced combustion (MIC) digestion according to the procedure described in previous work.[20] An anion exchange column (Metrosep A Supp 5, with 150 × 4 mm i.d.) and a guard column (Metrosep A Supp 4/5 Guard) were used. The suppressor column was periodically regenerated with water and 50 mmol L^{-1} H$_2$SO$_4$. A solution of 3.2 mmol L^{-1} Na$_2$CO$_3$ and 1 mmol L^{-1} NaHCO$_3$ was used as mobile phase at a flow rate of 0.7 mL min^{-1} as well as a dialysis cell composed of a cellulose triacetate membrane. Results were reported as NaCl content.

The determination of TAN in crude oil before and after sediment removal was performed following the method recommended by ASTM[21] using an automatic titrator (Titrando 836, Metrohm, Switzerland) equipped with a combined glass electrode for non-aqueous media. The same equipment was used for water content determination according to ASTM D 4377 standard test method, using a platinum electrode.[22]

For the N and S determinations, a specific analyzer (model 9000 series, Antek Instruments, USA) was used and the procedures were carried out according to ASTM D 4629[23] and ASTM D 5453[24] methods, respectively. Samples obtained after sediment removal using microwave radiation were homogenized, dissolved in toluene and injected into the high temperature combustion tube using a syringe.

The determination of density and viscosity was performed according to ASTM D 7042[25] method, using a Stabinger viscometer (model SVM 3000, Anton Paar, Austria). Samples obtained after sediment removal were homogenized and introduced into the measuring cells with controlled temperature.

Chemicals, Solutions and Samples

All reagents were of analytical grade. Purified water from a Milli-Q® system (Milipore, Billerica, USA) was used for sediment removal using microwave radiation and also to prepare all reagents, standard solutions and mobile phase. Stock standard solution of Cl (1000 mg L^{-1}) was prepared by dissolving NaCl (Merck, Darmstadt, Germany) in water and analytical standards were prepared by sequential dilution of this solution in water. Sodium carbonate (Merck) and NaHCO$_3$ (Merck) were used to prepare the mobile phase, and H$_2$SO$_4$ (Merck) was used to prepare the solution used for suppressor column regeneration. Glass materials were cleaned using toluene and water and further soaked in 20% (v v^{-1}) HNO$_3$ (Merck) and rinsed with water before use.

Water content was determined in the crude oil before and after sediment removal using Karl Fischer (two-component) reagent Composite 5 (Riedel-de Häen, Seelze, Germany). A mixture of toluene (Vetec Química Fina Ltda., Rio de Janeiro, Brazil) and methanol (Carlo Erba Reagents, Milan, Italy) (3 + 1) was used for sample dissolution. Toluene (Vetec) was also used for sediment determination in crude oil.

Standard solutions for S and N determination were prepared after dissolution of dibenzothiophen (DBT) (C$_{12}$H$_8$S, $\geq$ 98%, Merck) and pyridine (Merck) in toluene, respectively. Naturally emulsified heavy crude oil samples, identified as A, B and C, were used. Heavy crude oil A was selected for development and optimization of the proposed method that was subsequently applied for other samples.

Microwave-assisted Sediment Removal Method

Heavy crude oil emulsions were heated at 60 °C for 30 min and homogenized using a mechanical stirrer for 15 min. Subsequently, 20 g of sample were weighted directly into the quartz vessels and 20 mL of water were added. Five glass spheres (ø = 5 mm) were also added to avoid sample projection during the microwave heating. The proposed microwave heating procedure was performed using operational parameters described in Table 1.

Table 1: Microwave heating program used for sediment removal from crude oil emulsion

Parameter	
Sample mass / g	20
Water volume / mL	20
Power / W	1400
Ramp / min	5
Heating time / min	5
Cooling / min	20
Maximum temperature / °C	260
Maximum pressure / bar	70

After the end of the heating program, the oil phase was transferred to a polypropylene vessel using a syringe. Then, water containing the sediments was removed from quartz vessels and also transferred to vessels. Before characterization, crude oil was centrifuged (7000 rpm) in order to improve the separation of free water droplets, according to the procedure described in a previous work.[18]

Heavy crude oil was characterized by the determination of sediments, salt (as NaCl), water content, viscosity, density, N and S. After each microwave run, vessels were cleaned using, firstly, toluene for removing oil residues, then with concentrated nitric acid by heating in microwave oven for 10 min and rinsed with water.

Determination of Sediments in Crude Oil Before and After Microwave Irradiation

Sediment content in heavy crude oil was determined according to ASTM D 4807 method.[12] Briefly, the procedure consists of filtering a known amount of crude oil with toluene at 90 °C through a 0.45 μm pore size nylon membrane. The weight of membrane is recorded (after drying at 105 °C) before and after crude oil filtration. The mass of solids (in percentage) is calculated by the difference of weight of membrane and related to the amount of filtered crude oil. This procedure was applied before and after sediment removal to evaluate the efficiency of the proposed method for sediment removal from crude oil.

The water phase obtained after microwave-assisted sediment removal method was also filtered through a 0.45 µm pore size nylon membrane for the solid content determination. The membrane was washed with hot toluene to remove oil residues from solid particles, and the quantification was performed by weighing, as it was performed for crude oil. Additionally, water phase was also analyzed for determination of total solid content. In this case, the determination was performed following a procedure commonly used for total solid determination in water.[26] Water phase was transferred to a previously dried and weighed platinum crucible, left in an oven until dry at 80 °C and after at 105 °C for 1 h. After cooling, the weight of crucible plus solids was recorded, and total solid content was calculated considering the mass of crude oil from respective water phase that was submitted to microwave heating procedure.

RESULTS AND DISCUSSION

Characterization of the Crude Oil Emulsions

Before optimizing the microwave-assisted sediment removal method, crude oil emulsions were characterized through the determination of sediments and water contents using ASTM D 4807 and D 4377 methods, respectively Additionally, the determination of Cl (reported as NaCl salt) using MIC was also performed.

Results obtained for sediments, water and salt for samples A, B and C before the proposed microwave-assisted method in closed vessels are shown in Table 2.

Table 2: Sample characterization before proposed microwave-assisted method in closed vessels (n = 3)

Parameter	A		
Water content / %	22.0 ± 0.6	30.8 ± 0.5	3.20 ± 0.05
Sediment content / %	5.15 ± 0.20	3.49 ± 0.15	1.13 ± 0.05
Salt / (µg g⁻')	41700 ± 3300	19274 ± 670	12128 -± 1156

Sediments and salt content in crude oil emulsion samples were considered relatively high, justifying the interest in developing a unique method that could be suitable for both sediment and salt extractions using the same procedure. It is important to mention that TAN could not be determined in all the emulsions (before the proposed procedure) due to interferences during titration using the ASTM D 664 method. The titration end points were not well defined and no repeatability between results could be obtained during successive titrations. These interferences can be attributed to inorganic salts or sediments present in crude oil emulsions.[27]Additionally, parameters such as density and viscosity could not be determined due to the presence of sediments and water that interfere in the analysis. The S and N determinations were not carried out before sediment removal because the sample is injected into a specific analyzer using a syringe and the presence of solid particles could block the syringe needle.[23,24]

Optimization of the Microwave Heating Program for Sediment Removal

Microwave radiation in closed vessels has been already applied for salt extraction from heavy crude oil, with addition of water. During the microwave heating, water refluxes inside the quartz vessels and allows washing of the crude oil, transferring salt to the water phase.[8] In case of sediments, it would be expected that water could carry the solid particles from the crude oil to the water. Therefore, preliminary studies were performed for sample A using similar conditions employed in previous work for salt extraction (800 W power; 30 min of irradiation).[8] At the end of the heating program, brown solid particles were observed in the water phase (on the bottom of quartz vessels), showing that microwave radiation could be feasible for sediment removal from heavy crude oil. After this initial test, microwave heating program was optimized, using 5 g of heavy crude oil and 20 mL of water. A three level full factorial design of two variables was used, as shown in Table 3. Experiments were performed in triplicate in random order for combination of levels.

Table 3: Full factorial experimental design used for optimizing the microwave heating program for sediment removal from crude oil emulsions

Run	X_1	X_2
1	-	-
2	0	-
3	+	-
4	-	0
5	0	0
6	+	0
7	-	+
8	0	+
9	+	+

X_1: heating time (5. 15 and 30 min): X_2: microwave power (300. 800 and 1400 W)

Results from the factorial design were processed by analysis of variance (ANOVA) with confidence level of 95%. Sediments were determined in aliquots of crude oil obtained after microwave heating according to ASTM D 4807-05 method.[12]

Sediment removal was more efficient (Table 4) when microwave power was set at 1400 W. In this case, using only 5 min of heating at maximum temperature, the sediment content in sample A was reduced from 5.15 ± 0.20 to 2.9 ± 0.24% (43.7% of removal efficiency). Therefore, this heating program was chosen for subsequent tests.

(NaCl) in dissolved solids was 64.3 ± 4.0% (m m^{-1}). According to these evidences, it could be expected that part of solids determined in the crude oil using the ASTM D 4807-05 method are composed of chloride salts.

Characterization of the Crude Oil after Sediment Removal Method

Crude oils obtained after the proposed sediment removal method were characterized for comparison of characteristics previously reported, before sediment removal (sediments, water and salt content). Additionally, other parameters, such as TAN, density, API gravity, kinematic and dynamic viscosity were determined after sediment removal and separation of water. The results obtained for samples A, B and C are shown in Table 5.

Table 5: Characterization of the crude oil samples after sediment removal using the proposed microwave-assisted method in closed vessels (n = 3)

Parameter	A	B	C
API gravity	18.0	19.6	21.0
Kinematic viscosity / (mm^2 s'). 80 °C	74.825 ± 0.269	87.401 ± 0.380	46.211 ± 0.401
Dynamic viscosity / (mPa s), 80 °C	67.559 ± 0.412	78.072 ± 0.338	36.475 ± 0.390
Water content I %	0.20 ± 0.01	0.45 ± 0.02	0.12 ± 0.01
Sediment content / %	<0.10	< 0.10	<0.10
Salt / (µg g^{-1})	< 6.0	9.54 ± 0.67	21.9± 1.6
TAN / (mg KOH g^{-1})	0.46 ± 0.02	0.84 ± 0.06	0.45 ± 0.01
Nitrogen / (µg g^{-1})	2160=97	1712 ± 53	3496 ± 89
Sulfur / (µg g')	3839 ± 201	3236 ± 65	4979 ± 134

It is possible to observe that the water content in crude oil samples was lower than 0.50%, giving an efficiency of separation of water better than 96%. This result indicates the good performance of microwave radiation for water separation from crude oil emulsions in agreement with previous work.[18] In the same way, the sediments and salt content were lower than 0.10% and 22 µg g[-1], respectively. Several parameters could be determined in the crude oil samples after sediment removal method without interferences, such as the determination of API gravity, kinematic and dynamic viscosity, S and N. In case of TAN, the characteristic titration curves and well defined titration end points were obtained during titration of the crude oil after sediment removal.

The proposed method could be considered advantageous in comparison with other methods used for sediment removal and subsequent crude oil characterization.[27,28] For example, the simple distillation procedure allows removing only water, but not chloride salts.[28] Using microwave radiation in closed vessels, it was possible to remove sediments and salt, as well as to separate water from heavy crude oil emulsion, in a single method, in a relatively short time and for eight samples simultaneously.

CONCLUSIONS

The efficiency of the proposed method using microwave radiation in closed vessels for sediment removal was better than 95%. Moreover, the salt and water contents were drastically reduced allowing the determination of several important parameters for crude oil characterization without interferences. Up to eight samples were simultaneously processed in a little time, resulting in a good sample throughput, although 5 extraction steps were necessary for almost complete sediment removal. The sediment removal process could be performed using only water, avoiding the use of toxic reagents or demulsifiers, an important aspect related to the green chemistry recommendations. Additionally, the use of microwave-assisted method is relatively easy to be performed and could be suggested for routine sediment removal from heavy crude oil for subsequent characterization.

ACKNOWLEDGEMENTS

The authors are grateful to CNPq and CAPES for supporting this study and also to CENPES/Petrobras for support and donation of samples.

REFERENCES

1. Gafonova, O. V.; Yarranton, H. W.; *J. Colloid Interface Sci.* 2001,*241*,469.

2. Sullivan, A. P.; Kilpatrick, P. K.; *Ind. Eng. Chem. Res.* 2002,*41*,3389.

3. Poindexter, M. K.; Chuai, S.; Marble, R. A.; Marsh, S. C.; *Energy Fuel* 2005,*19*,1346.

4. Poindexter, M. K.; Marsh, S. C.; *Energy Fuel* 2009,*23*,1258.

5. Cloud, R. W.; Marsh, S. C.; Ramsey, B. L.; Pultz, R. A.; Poindexter, M. K.; *Energy Fuel* 2007,*21*,1350.

6. El Gamal, M.; Mohamed, A. M. O.; Zekri, A. Y.; *J. Petrol. Sci. Eng.* 2005,*46*,209.

7. Mohamed, A. M. O.; El Gamal, M.; Zekri, A. Y.; *J. Petrol. Sci. Eng.* 2003,*40*,177.

8. Moraes, D. P.; Antes, F. G.; Pereira, J. S. F.; dos Santos, M. D. F. P.; Guimaraes, R. C. L.; Barin, J. S.; Mesko, M. F.; Paniz, J. N. G.; Flores, E. M. M.; *Energy Fuel* 2010,*24*,2227.

9. Al-Otaibi, M.; Elkamel, A.; Al-Sahhaf, T.; Ahmed, A. S.; *Chem. Eng. Commun.* 2003,*190*,65.

10. Behin, J.; Aghajari, M.; *Sep. Purif. Technol.* 2008, *64*, 48.

11. Fortuny, M.; Oliveira, C. B. Z.; Melo, R. L. F. V.; Nele, M.; Coutinho, R. C. C.; Santos, A. F.; *Energy Fuel* 2007,*21*,1358.

12. ASTM D 4807-05: *Standard Test Method for Sediment in Crude Oil by Membrane Filtration*, Annual Book of ASTM Standards, USA, 2005.

13. Lee, R. F.; *Spill Sci. Technol. Bull.* 1999,*5*,117.

14. Cloud, R. W.; Marsh, S. C.; Linares-Samaniego, S.; Poindexter, M. K.; *Energy Fuel* 2010,*24*,2376.

15. ASTM D 6470: *Standard Test Method for Salt in Crude Oils (Potentiometric Method)*, Annual Book of ASTM Standards, USA, 2004.

16. Chan, C. C.; Chen, Y. C.; *Sep. Sci. Technol.* 2002,*37*,3407.

17. Fang, C. S.; Chang, B. K. L.; Lai, P. M. C.; Klaila, W. J.; *Chem. Eng. Commun.* 1988,*73*,227.

18. Diehl, L. O.; Moraes, D. P.; Antes, F. G.; Pereira, J. S. F.; Santos, M. D. F. P.; Guimaraes, R. C. L.; Paniz, J. N. G.; Flores, E. M. M.; *Sep. Sci. Technol.* 2011,*46*,1358.

19. Pereira, J. S. F.; Diehl, L. O.; Duarte, F. A.; Santos, M. F. P.; Guimaraes, R. C. L.; Dressler, V. L.; Flores, E. M. M.; *J.Chromatogr., A* 2008,*1213*,249.

20. Pereira, J. S. F.; Mello, P. A.; Moraes, D. P.; Duarte, F. A.; Dressler, V. L.; Knapp, G.; Flores, E. M. M.;*Spectrochim. Acta, Part B* 2009,*64*,554.

21. ASTM D 664-06: *Standard Test Method for Acid Number of Petroleum Products by Potentiometric Titration*, Annual Book of ASTM Standards, USA, 2006.

22. ASTM D 4377-00: *Standard Test Method for Water in Crude Oils by Potentiometric Karl Fischer Titration*, Annual Book of ASTM Standards, USA, 2000.

23. ASTM D 4629-02: *Standard Test Method for Trace Nitrogen in Liquid Petroleum Hydrocarbons by Syringe/Inlet Oxidative Combustion and Chemiluminescence Detection*, Annual Book of ASTM Standards, USA, 2002.

24. ASTM D 5453-06: *Standard Test Method for Determination of Total Sulfur in Light Hydrocarbons, Spark Ignition Engine Fuel, Diesel Engine Fuel, and Engine Oil By Ultraviolet Fluorescence*, Annual Book of ASTM Standards, USA, 2002.

25. ASTM D 7042-04: *Standard Test Methods for Dynamic Viscosityand Density of Liquids by Stabinger Viscometer (and the Calculation of Kinematic Viscosity)*, Annual Book of ASTM Standards, 2004.

26. Greenberg, A. E.; Clesceri, L. S.; Eaton, A. D.; *Standard Methods for the Examination of Waterand Wastewater*, 21st ed.; American Public Health Association: Washington, USA, 1992.

27. Santos, M. F. P.; Guimarães, R. C. L.; Gomes, L. M. B.; Camacho, C. F. B.; Trindade, F. F.; *Technical Report*; Petrobras: Rio de Janeiro, Brasil, 2006.

28. Petrobras Procedure N-2499: *Dehydration of Crude Oil by Distillation. SC-20, Técnicas Analíticas de Laboratório*, Rio de Janeiro, Brasil, 2003.

Chapter 7

Effects of Direct and Alternating Current on the Treatment of Oily Water in an Electroflocculation Process

A. A. Cerqueira, P. S. A. Souza; M. R. C. Marques

Instituto de Química, Laboratório de Tecnologia Ambiental, Universidade do Estado do Rio de Janeiro, UERJ, Rio de Janeiro - RJ, Brasil

ABSTRACT

In the direct current mode (DC), widely used in electro flocculation (EC), the formation of an impermeable oxide layer on the cathode causes the declining of the efficiency of this process. This disadvantage has been reduced by adopting alternating current (AC). In this study, the effects of AC and DC on operational parameters such as the removal of oils and greases (O&G), color and turbidity from oil-in-water (O/W) emulsions of the petroleum industry using aluminum electrodes were investigated. Removal efficiencies of 95%, 97% and 99% of O&G, color and turbidity with energy consumption of 0.280 kWh/m³ and electrode consumption of 0.12 g and 0.18 g were achieved at a current density of 3 A, operation time of 3 minutes and initial pH of 9.0 using AC and DC, respectively. In continuous flow tests performed with the same experimental conditions, the electrode consumption at times up to 60 minutes were 1.6 g and 3.4 g using AC and DC, respectively.

INTRODUCTION

Produced water, one of the main problems in the petroleum industry, is generated in increasing volumes from both old and new wells (Campos *et al*, 2005). This effluent represents 98% of all wastewater generated in the petroleum industry. This water contains complex organic and inorganic substances, such as salts, metals, dispersed oil and dissolved hydrocarbons and also presents high temperature and the absence of oxygen (Thomas, 2004).

Therefore, produced water is an important source of pollutants and, as the environmental laws have become stricter, the cost of its treatment has become increasingly high. This fact has led to many efforts to find more effective and less expensive ways to treat this water (Li *et al.*, 2009).

The treatment of produced water applies a primary treatment to separate the floatable oils from the water and emulsified oils. This treatment process usually involves retaining the oily wastewater in a holding tank, while allowing gravity separation of the oily material, which is subsequently skimmed from the wastewater surface. Meanwhile, a secondary treatment phase is then required to break the oil - water emulsion and separate the remaining oil from the water. Emulsions may be broken by chemical, physical, or electrical methods. Chemical methods are the most widely used in this treatment. However, these chemical methods present troublesome filtration processes, providing an incentive to explore other alternatives (Yang, 2007).

In recent years, electrocoagulation (EC) of synthetic or real oily water has been investigated by some researchers (Ruback and Saur, 1997; Dórea *et al.*, 2007; Tir and Mostefa, 2008; Canizares *et al.*, 2007; Benzadok*et al.*, 2008; Asselin *et al.*, 2008; Abdelwahaba *et al.*, 2009; Ramalho *et al.*, 2010; Lima *et al.*, 2009; Cerqueira *et al.*, 2011). During electrolysis, four main mechanisms generally occur simultaneously: (a) electrolytic reactions at the electrode surfaces, (b) formation of coagulant agents in the aqueous phase, (c) adsorption of soluble or colloidal pollutants by these agents and (d) removal by sedimentation or flotation. Aluminum is released from the anode (reaction 1) and hydrogen gas is formed at the cathode (reaction 2):

$$Al \rightarrow Al^{3+}_{(aq)} + 3e^- \quad (anode)$$

(1)

$$3H_2O + 3e^- \rightarrow 3/2H_2 + 3OH^- \ (cathode)$$

(2)

The aluminum chemical reaction in water is very complex, because aluminum is capable of forming several compounds, such, as $Al(OH)^{2+}$, $Al(OH)_2^+$, $Al_2(OH)_2^{4+}$, $Al(OH)_4^-$, as well as polymeric species like $Al_6(OH)_{15}^{3+}$, $Al_7(OH)_{17}^{4+}$, $Al_8(OH)_{20}^{4+}$, $Al_{13}O_4(OH)_{24}^{7+}$, $Al_{13}(OH)_{34}^{5+}$, which are finally transformed into $Al(OH)_3$ according to complex precipitation kinetics (Gurses et al., 2002; Rebhun and Larue, 1993).

The advantages of electrocoagulation include high particulate removal efficiency, a compact treatment facility, relatively low costs and the possibility of complete automation. This method is characterized by reduced sludge production, a minimum requirement of chemicals and easy operation (Chen, 2004; Vasudevan et al., 2011).

In the direct current (DC) technology, widely used in EC, the anode oxidation causes the formation of an impermeable oxide layer on the cathode, which increases the resistivity of the electrode. With the time, the efficiency of the EC process declines. This problem can be minimized by addition of sacrificial electrodes in a parallel configuration in the electrolytic cell (Yousuf et al., 2001; Cerqueira et al., 2009). Furthermore, to reduce the cathode passivation and to extend the lifetime of the electrodes, the direction of current at regular intervals of time can be manually inverted. Thus, cathode and anode can be switched periodically. However, many researchers have preferred the use of alternating current in the EC process. It is assumed that the cyclic energization between the anodes - cathode in an alternating current (AC) system simulates the manual reversion of polarity. It delays the cathode passivation and anode deterioration and thus ensures reasonable electrode life (Vasudevan et al., 2011).

Eyvas *et al.* (2009) investigated the effects of alternating current (AC) on dye removal from aqueous solutions by electrocoagulation (EC). An EC system with parallel-connected aluminum electrodes was operated in batch mode. Two different aqueous dye solutions were used. The experiments employing direct current (DC) were carried out using a DC power supply. The AC experiments were conducted using a rectangular wave, which is produced with an adjustable time relay connected to the output of a DC power supply. This current was called alternating pulse current (APC) in order to refer to the AC system in this study. Total organic carbon (TOC) and dye removal efficiencies were measured to assess treatment efficiency. The results of this study showed that high removal efficiencies of TOC and dye can be acquired in shorter operation times by using an APC system (5 min of operation time).

Keshmirizadeh *et al.* (2011) showed that electroflocculation (with Fe/Al electrodes) could be applied in the treatment of industrial effluents containing Cr^{6+}. The alternating pulse current (APC) mode was found to be more efficient than the DC mode with a lower anode over-voltage, slower anode polarization and passivity, and lower tank voltage. The operating time was 3 - 25% less when the APC mode was used, based on an initial Cr^{6+} concentration of 50 - 1000 mg/L, respectively. Because of the reduction in operating time, less power (or energy) is consumed, which makes the APC mode more cost effective. Application of APC eliminates uneven wear (dissolution) of electrodes. Typically, the anode material dissolves and electroreduction products stick to the cathode when the DC mode is used. In this study, water recovery was found to range from 0.7 to 0.92, based on initial Cr^{6+} concentrations from 50 to 1000 mg/L, respectively.

Vasudevan *et al.* (2011) investigated the effects of AC and DC on the removal of cadmium from water using aluminum electrodes as anode and cathode. The results showed that removal efficiencies of 97.5 and 96.2% with energy consumptions of 0.454 and 1.002 kWh m^3 were achieved at a current density of 0.2 A/dm^2 and pH of 7.0 using aluminum alloy electrodes and AC and DC, respectively. The aluminum hydroxide generated in the cell reduced the cadmium concentration in water to less than 0.005 mg/L and made it suitable for drinking. The results indicate that the process can be scaled up to higher capacity.

The electrocoagulation with the AC process was compared to the chemical coagulation process for the treatment of oily waste generated by the petroleum industry (Cerqueira *et al.*, 2011). From the results, one may conclude that this EC process was effective for the effluent studied, while chemical coagulation was not successful.

The main objective of this study was to investigate the effects of AC and DC treatment using aluminum electrodes as anode and cathode in order to evaluate the decrease of turbidity, color and oil and grease content (O&G) from a synthetic oil/water (O/W) emulsion. Important electrochemical factors were investigated such as: initial pH, current density, distance between electrodes and electrocoagulation time.

EXPERIMENTAL

Synthetic Oil/Water (O/W) Emulsion

To simulate produced water, a synthetic oil/water (O/W) emulsion was prepared in a 2 L becker containing 1 g of crude oil (from the Campos Basin, Rio de Janeiro state, Brazil, with density 0.89 g/L and 28 °API), 0.1 g/L of the emulsifiers SP-60® and TW-60® (1:1 ratio - Oxiteno Corp.) and 0.9 L of distilled water salinized with synthetic sea salt (60 g/L - Coralife Corp.). This mixture was then subjected to vigorous mechanical stirring at 10,000 rpm (Wigen Hauser D-500 homogenizer) for 10 minutes to form a stable O/W emulsion

Table 1: Characteristics of simulated oily wastewater

Parameter	Values
Color ($A_{400\ nm}$)	2.3 — 2.9
Turbidity (NTU)	4100 — 4750
pH	7.9 — 9
Conductivity (mS/cm)	98 —100.7
O&G (mg/L)	650 - 690

Electrocoagulation Experimental Set-Up

Batch Reactor

A monopolar electrode with two pairs of aluminum plates (10 cm x 5 cm x 0.3 cm) was placed vertically in a 1.5 L becker. The interelectrode distance was variable at 0.5 to 2.0 cm. The runs were performed using both AC and DC sources and at ambient temperature (25 °C). The pH was adjusted to the desired value using NaOH or H_2SO_4. All tests were performed in triplicate and kept under stirring at 200 rpm. The weight loss of the electrodes was evaluated after cleaning, drying and weighing each electrode in order to assess the best operational conditions. Between the tests, the electrolytic cell (including the electrodes) was cleaned with 5% (v/v) hydrochloric acid solution for at least 15 min and then rubbed with a sponge and rinsed with tap water.

Continuous Flow Reactor

In a 5 L supply tank, kept under continuous agitation to assure that the effluent was emulsified, a peristaltic pump (Exata, EX 20 SV) was connected to allow flow control with pre-determined times to feed the electrolytic cell. This cell consisted of a 1.8 L glass tank with four deflectors in which monopolar electrodes were vertically inserted with 4 pairs of aluminum plates (10 cm x 5 cm x 0.3 cm) and separated by spacers 1.0 cm thick each. The electrode mass consumption was determined by cleaning, drying and weighing each electrode before and after each test. The time interval samples were taken from the reactor at 10th, 20th, 30th, 40th, 50th and 60thminutes of the treatment times. All assays were performed in triplicate. Both AC and DC current were used. In the DC electroflocculation tests, the polarity was reversed every 5 minutes (Figures 1 and 2). After 30 min, the effectiveness of each parameter was determined by the differences in turbidity, color and O&G content between treated and untreated emulsions. Before the analysis, the treated solutions remained for 30 min without stirring for separation of the oily material.

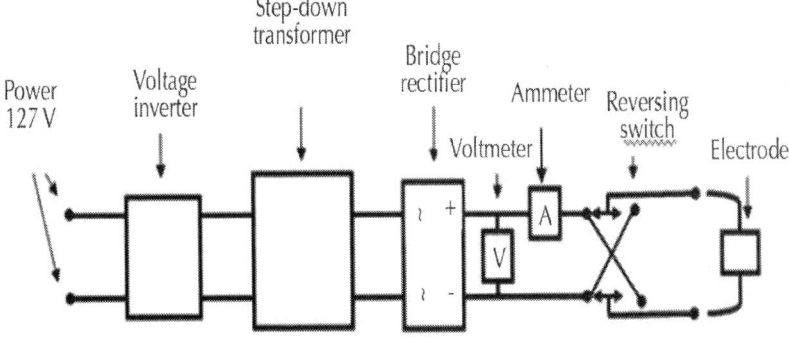

Figure 1: Schematic diagram of the experimental DC electroflocculation unit.

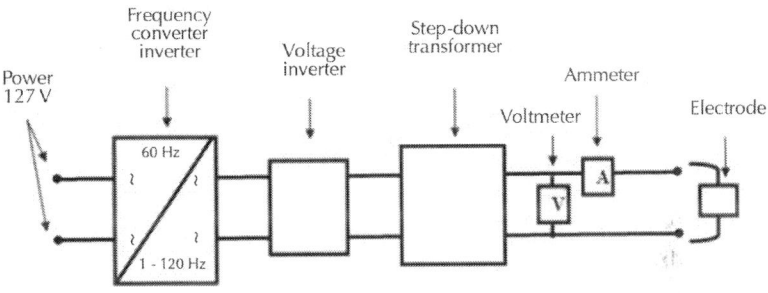

Figure 2: Schematic diagram of the experimental AC electroflocculation unit.

Description of Alternating Current and Direct Current Electrocoagulation

All tests, batch or continuous flow reactor, were conducted with two electrolytic units: DC and variable frequency AC. The DC electroflocculation unit (15 V) was composed of a voltage inverter plugged into a standard AC outlet (127 V/60 Hz), then connected in line to a step-down transformer feeding a bridge rectifier, responsible for providing DC to the electrodes through a polarity reversing switch,

Table 2: Effect of current density (CD) on the removal efficiency of O&G, energy and electrode consumption using AC and DC. Initial O&G, 690 mg/L; conductivity, 100, 7 uS/cm and initial pH, 9.0

CD (A)	AC			DC		
	Removal efficiency (%)	Energy consumption (kWh/m³)	Electrode consumption (kg/m³)	Removal efficiency (%)	Energy consumption (kWh/m³)	Electrode consumption (kg/m³)
1	81.0±2.0	0.09±0.01	0.07±0.01	75.8±3.1	0.07±0.01	0.11±0.01
2	91.0±1.7	0.18±0.01	0.09±0.01	90.3±2.1	0.17±0.01	0.15±0.02
3	94.3±0.6	0.28±0.00	0.12±0.10	94.1±0.6	0.28±0.02	0.18±0.10
4	92.7±0.6	0.38±0.04	0.15±0.20	93.5±0.6	0.46±0.03	0.22±0.20
5	93.7±0.6	0.48±0.10	0.18±0.20	94.2±0.6	0.69±0.10	0.26±0.30

In this study, the high conductivity of the O/W emulsion (around 100.7 mS/cm) caused a very low voltage (~ 2.0 V) for a current density of 3 A in both technologies. The increase in conductivity of the effluent favors electrical current conduction, reducing the voltage between the electrodes, and consequently requiring less energy for the electrolytic process. According to Daneshvar *et al.*, (2004), higher electrical current density increases the dissolution of the anode, producing a larger quantity of coagulation agent generated in a fixed time. The presence of coagulant in solution contributes to an increase in oil removal efficiency by flotation of the oil droplets, reduction of electrostatic repulsion between the air bubble and oil droplets and an increase of oil droplet hydrophobicity. Moreover, the rate of generation of bubbles increases and the bubble size decreases. These effects are favorable for destabilization of the emulsion. The effect of current density on the final removal efficiency of chemical oxygen demand (COD) upon increasing the applied power was also observed by Bensadok *et al.* (2008), Khemis *et al.* (2005) and Koby*et al.* (2006).

In the present investigation, a current density around 3 A seems to be enough for a better electrolytic flocculation for both AC or DC sources and consequently a maximum efficiency of removal (99% turbidity, 97% color and 94% O&G).

Since increasing the applied current density means a higher energy consumption, the current density of 3 amperes (A) was selected for the next experiments using the DC source, because increasing the density did not improve the removal efficiencies of O&G, color and turbidity.

Influence of Initial pH

It has been established that the initial pH is an important parameter in determining the performance of the electrocoagulation process. In order to examine the effect of pH on the removal efficiencies of turbidity, color and O&G, the pH was varied between 4 and 9. The maximum removal was at pH above 6 in the electrocoagulation process using the DC source: 96% O&G, 99% turbidity and 99% color (Figure 3). These results are similar to those obtained in previous studies using an AC source by Cerqueira *et al.*, (2011). However, in this case, at pH 5 the efficiency was lower than with the DC source: 61% O&G, 6% turbidity and 37% color. In both processes (AC or DC) the removal remained unchanged until pH 9. Experiments at pH above 9 were not performed because it is known that the flocs of aluminum hydroxide are less reactive and the flocculation is less effective. Nevertheless, there was good removal efficiency at neutral pH. This result is an advantage of electrolysis for the treatment of produced water, considering that this type of effluent generally has pH between 7 and 9 (Queiroz et al., 1996).

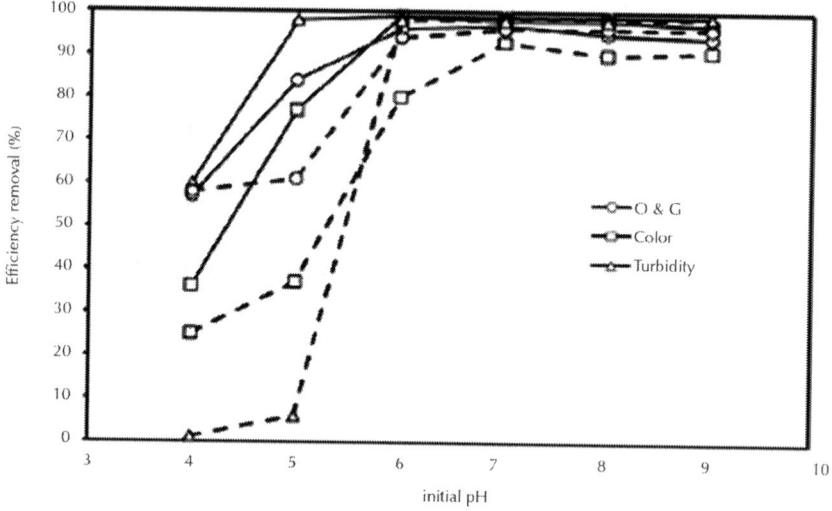

Figure 3: Effect of initial pH on color, turbidity and O&G with the DC source (—) and the AC source (---). Conditions: current intensity 3A, time lapse: 3 min., interelectrode distance: 0.5 cm, oil concentration: 690 mg/L, conductivity: 100.7 uS/cm, initial pH: 9.0, color: 2.9 abs, turbidity: 4750 NTU, temperature: 25 °C and emulsifier: 100 mg/L.

A comparison of the turbidity test results showed that, with pHs of 4 and 5, removal occurred only slightly with the AC source. For the DC source at initial pH 4, the removal was 60% and at pH 5 it was 98%. One hypothesis to explain this difference is that, with AC, at the beginning of the aluminum oxidation there is no formation of the minimum amount of coagulant required to treat the effluent in this pH range. In the case of DC, the electrode oxidation is probably higher since, even when the pH is not optimal for coagulation, there is greater formation of flocculating factors. From pH 6 to 9, both AC and DC behave alike in the removal parameters analyzed, since in this pH range there is formation of the primary flocculating agent Al (OH) $_3$.

Therefore, because the pH of this oily water was 9, it was selected for the next experiments.

Effect of Interelectrode Distance

The interelectrode distance is an important variable in order to optimize the operating costs of electrolysis systems. Researchers report that when the conductivity of the effluent is high, a larger spacing between the electrodes is possible. On the other hand, when conductivity is low, the spacing should be smaller (Crespilho and Rezende, 2004).

In this study, the interelectrode distance (4 aluminum electrodes in the cell) was varied from d = 0.5 to 2.0 cm while the other factors remained unmodified. In Figure 4, the removal efficiencies for all parameters were above 90% for the DC source, indicating that the inter-electrode distance did not greatly affect the performance. These results are similar to those obtained in the electrocoagulation using the AC source under the same conditions by Cerqueira et al. (2011).

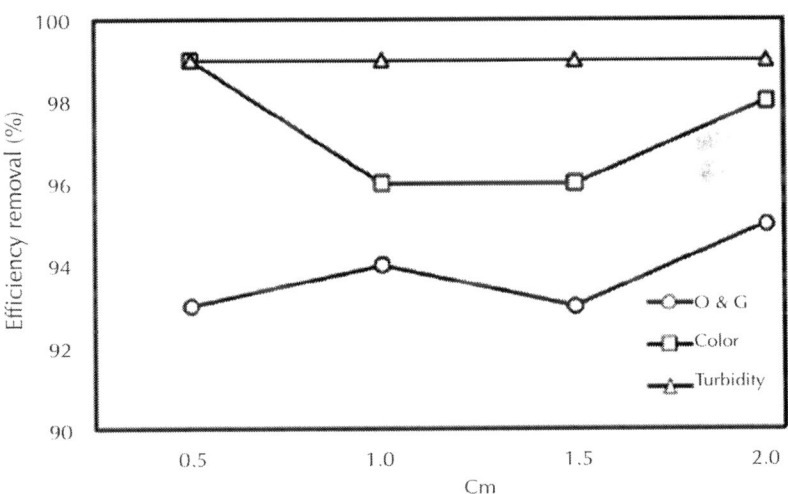

Figure 4: Effect of interelectrode distance on removing color, turbidity and O&G. Experimental conditions: initial pH 9, time 3 min., Current intensity 3 A, DC source, oil concentration: 690 mg/L, conductivity: 100.7 uS/cm, initial pH: 9.0, color: 2.9 abs, turbidity: 4750 NTU, temperature: 25 °C and emulsifier: 100 mg/L.

When the distance between electrodes increases, the energy consumption also increases due to the higher resistivity of the solution (Den and Huang, 2005). The results in Figure 4 showed no improvement in the efficiency of reduction of the parameters for distances between the electrodes greater than 0.5 cm using a current density of 3 A, indicating that this is the optimal distance since the use of larger distances would involve greater energy consumption (0.5 cm = 1.95V; 2.0 cm = 3.00V).

Effect of Time

Regarding the time, experiments were carried out at pH 9 for an initial O&G concentration of 690 mg L^{-1} in the presence of 60 g L^{-1} NaCl. The variation in the O&G removal considering the electrolysis time for oily wastewater at different current densities revealed that, using higher current densities, the oil removal is faster. The direct influence of the electrical current density on the time and the kinetics of removal of pollutants was also observed in previous studies (Chen et al., 2000; Kumar et al. 2004; Mouedhen et al., 2008).

The results presented in Figure 5 show the effect of variation of the electrolysis time (1 to 5 min) on the efficiency of O&G, color and turbidity removals at a current density of 3 A with the DC source. It appears that turbidity, color and O&G exhibit regular variations with a more continuous increase with time reaching a constant value. This behavior may be due to destabilization of the emulsion.

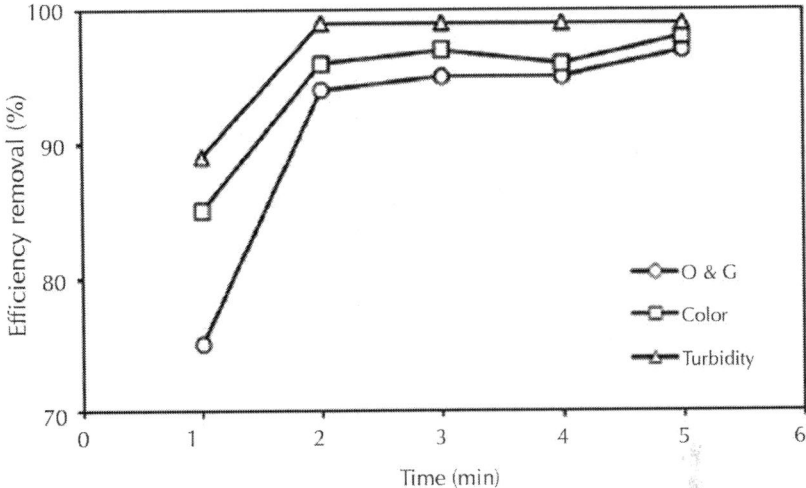

Figure 5: Effect of variation of time on removing color, turbidity and O&G. Experimental conditions: initial pH 9; current intensity 3 A, interelectrode distance 0.5 cm, DC source, oil concentration: 690 mg/L, conductivity: 100.7 uS/cm, color: 2.9 abs, turbidity: 4750 NTU, temperature: 25 °C and emulsifier: 100 mg/L.

The first 1 min of the electrocoagulation using the DC source gave a considerable removal of turbidity and color (above 85%) of the initial emulsion. The increase in the electrolysis time gave better results for the removal efficiencies reaching 99% turbidity, 97% color and 95% O&G, after just 2 min. However, with an increase of the electrolysis time, the removal efficiency remains constant and the emulsion becomes visually very clear. These results are similar to those obtained by Cerqueira et al. (2011) when an AC source was used under the same conditions (99% turbidity, 95% color and 96% O&G).

Electrolytic Treatment in Continuous Flow

Experiments were carried out in a continuous flow unit of 5 L capacity and contact times of 2, 4, 6, 8, 10 and 12 minutes per liter. After verifying the viability of the wastewater treatment by the electrolytic process with alternating current and direct current at low residence times in the system, the practicability of operation under continuous flow was checked. Figure 6 shows the efficiency of electrolysis as a function of time for AC and DC.

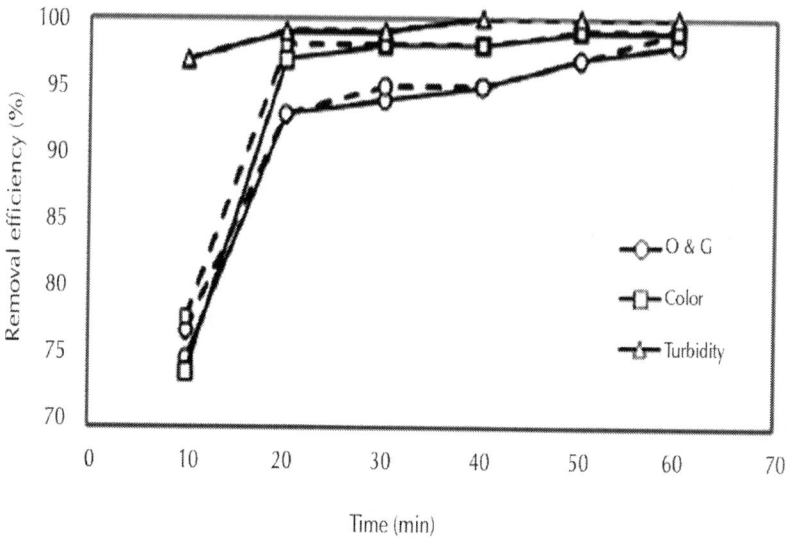

Figure 6: Removal efficiency of the indicated parameters after electroflocculation with DC (—) and AC (---). Conditions: initial pH 9, interelectrode distance 1 cm and current intensity 3 A, oil concentration: 690 mg/L, conductivity: 100.7 uS/cm, color: 2.9 abs, turbidity: 4750 NTU, temperature: 25 °C and emulsifier: 100 mg/L. Each test performed with 5 L of effluent.

A similarity in the removal was observed under the same conditions of AC and DC. Twenty minutes of electrolysis was enough for the removal with efficiencies above 90%. After this period, the further removal of an additional 5% is not justified by the higher consumption of electricity and electrodes.

Figure 7 shows the aluminum electrode weight loss by oxidation using the alternating current and direct current sources as a function of time in the continuous flow electrolysis.

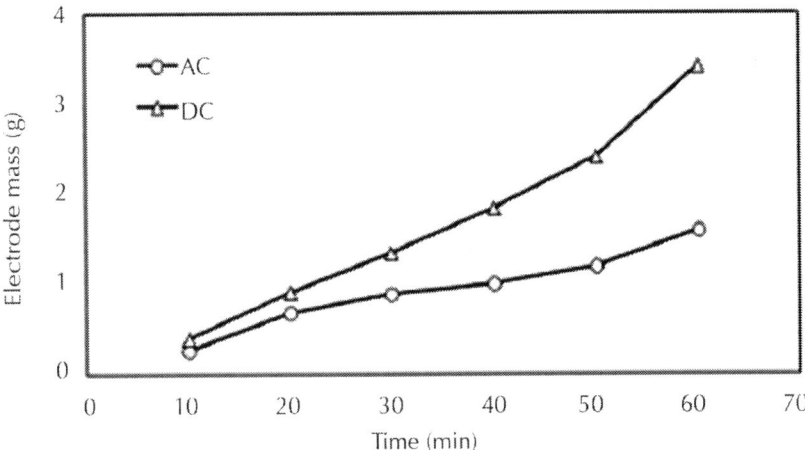

Figure 7: Aluminum electrode weight loss by oxidation by AC and DC for different electroflocculation times in continuous flow electrolysis.

After 10 minutes of AC electrolysis, the consumption of the aluminum electrode was 0.27 g and the consumption in DC was 0.4 g, whereas in 60 minutes of electrolysis the consumptions were 1.6 g and 3.4 g in AC and DC, respectively. Therefore, DC consumes the electrode much faster, meaning that AC performs better.

One hypothesis for the higher electrode consumption by DC in relation to AC is that, because DC current flows in one direction, it may cause irregular wear on the electrode plates due to the oxidation at the same preferential points of the electrode. In the case of AC, the reversal of current probably wears the electrodes more uniformly and allows a longer life.

CONCLUSIONS

In this study, the efficiency of the electroflocculation process applied to the treatment of oily wastewater emulsion was investigated with alternating current and direct current.

It was observed that the electroflocculation treatment achieves a fast and effective removal of turbidity, color and oil and greases. The treatment efficiency was found to be a function of the initial pH, interelectrode distance, applied current density and electrolysis time under the optimal values of the process parameters.

The results showed removal efficiencies of 94%, 97% and 99% of O&G, color and turbidity with an energy consumption of 0.280 kWh/m^3 with electrode consumption of 0.12 g and 0.18 g could be achieved at a current density of 3 A with an operation time of 3 minutes and initial pH of 9.0 using AC and DC, respectively.

Continuous flow tests performed under the same experimental conditions showed that the consumptions of electrodes with time up to 60 minutes of electroflocculation were 1.6 g and 3.4 g using AC and DC, respectively.

Therefore, the results indicated that the use of the AC mode of electroflocculation promoted a lower electrode consumption as compared to using DC under all conditions tested. This technique thus seems to be a promising alternative for the treatment of oil-in-water (O/W) emulsions of the petroleum industry.

ACKNOWLEDGEMENTS

We thank the Foundation for the Coordination for Improvement of Higher Education Personnel (CAPES), National Council for Scientific and Technological Development (CNPq) and the Research Foundation of the State of Rio de Janeiro (FAPERJ) for financial support. This study is part of the project INOG (Brazilian National Institute of oil and gas).

REFERENCES

1. Abdelwahaba, O., Aminb, N. K., El-Ashtoukhyb, E. Z., Electrochemical removal of phenol from oil refinery wastewater. J. Hazard. Mat., 163, p. 711-716 (2009).

2. APHA-AWWA, WPCF, Standard Methods for the Examination of Water and Wastewater. 21st Ed., American Public Health Association, Washington, DC (2005).

3. Asselin, M., Drogui, P., Brar, S. K., Benmoussa, H., Blais, J., Organics removal in oily bilgewater by electrocoagulation process. J. Hazard. Mat., 15, p. 1446-455 (2008).

4. Bensadok, K., Benammar, S., Lapicque, F., Nezzal, G., Electrocoagulation of cutting oil emulsion using aluminum plate electrodes. J. Hazard. Mat., 152, p. 423-430 (2008).

5. Campos, A. L. O., Rabelo, T. S., Santos, R. O., Melo, R. F. L. V., Cleaner production in the oil industry: The case of the water produced in the field of Carmópolis/SE. 23, ABES (2005).

6. Cañizares, P., Martínez, F., Lobato, Rodrigo, J. M. A., Break-up of oil-in-water emulsions by electrochemical techniques. J. Hazard. Mat., 145, p. 233-240 (2007).

7. Cerqueira, A. A., Marques, M. R. C., Russo, C., Evaluation of alternating current electrolytic process for treating produced water. Quim. Nova, 34, p. 59-63 (2011).

8. Cerqueira, A. A., Russo, C., Marques, M. R. C., Electroflocculation for textile wastewater treatment. Braz. J. Chem. Eng., 26, 4, p. 659-668 (2009).

9. Chen, G., Electrochemical technologies in wastewater treatment. Sep. Purif. Technol., 3, p. 11-41 (2004).

10. Chen, G., Chen, X., Yue, P. L., Electrocoagulation and electroflotation of restaurant wastewater. J. Environ. Eng., 126, p. 858-863 (2000).

11. Crespilho, F. N. and Rezende, M. O. O., Electroflotation: Principles and Applications. São Carlos, Rima (2004).

12. Daneshvar, N., Sorkhabi, H. A., Kasiri, M. B., Decolorization of dye solution containing Acid Red 14 by electrocoagulation with a comparative investigation of different electrode connections. J. Hazard. Mat., 112, p. 55-62 (2004).

13. Den, W. and Huang, C., Electrocoagulation for removal of silica nano-particles from chemical - mechanical-planarization wastewater. Colloids Surfaces A: Physicochem. Eng. Aspects, 254, p. 381-389 (2005).

14. Dórea, H. S., Bispo, J. R. L., Aragão, K. A. S., Cunha, B. B., Navickiene, S., Alves, J. P. H., Romão, L. P. C., Garcia, C. A. B., Analysis of BTEX, PAHs and metals in the oilfield produced water in the state of Sergipe, Brazil. Microchem. J., 85, p. 234-238 (2007).

15. Eyvaz, M., Kirlaroglu, M., Aktas, T. S., Yuksel. E., The effects of alternating current electrocoagulation on dye removal from

aqueous solutions. Chemical Engineering Journal, 153, p. 16-22 (2009).

16. Gürses, A., Yalçin, M., Doğar, C., Electrocoagulation of some reactive dyes: A statistical investigation of some electrochemical variables. Waste Manag., 22, p. 491-499 (2002).

17. Keshmirizadeha, E., Yousefia, S., Rofouei, M. K., An investigation on the new operational parameter effective in Cr (VI) removal efficiency: A study on electrocoagulation by alternating pulse current. J. Haz. Mat., 190, p. 119-124 (2011).

18. Khemis, M., Tanguy, G., Leclerc, J. P. G., Valentin, Lapicque, F., Electrocoagulation for the treatment of oil suspensions: Relation between the rates of electrode reactions and the efficiency of waste removal. Proc. Saf. Environ. Protect. 83, p. 50-57 (2005)

19. Kobya, M., Hiz, H., Senturk, E., Aydiner, C., Demirbas, E., Treatment of potato chips manufacturing wastewater by electrocoagulation. Desalination, 190, p. 201-21 (2006).

20. Kumar, P. R., Chaudhari, S., Khilar, K. C., Mahajan, S. P., Removal of arsenic from water by electrocoagulation. Chemosphere, 55, p. 1245-1252 (2004).

21. Li, G., Guo, S., Li, F., Treatment of oilfield produced water by anaerobic process coupled with micro-electrolysis. J. Environ. Science, 22, p. 1875-1882 (2009).

22. Lima, R. M. G., Wildhagen, G. R. S., Cunha, J. W. S. D., Afonso, J. C., Removal of ammonium ion from produced waters in petroleum offshore exploitation by a batch single-stage electrolytic process. J. Hazard. Mat., 161, p. 1560-1564 (2009).

23. Mouedhen, G., Feki, De Petris Wery, M., Ayed, H. F., Behavior of aluminum electrodes in electrocoagulation process. J. Haz. Mat., 150, 1, p. 124-135 (2008).

24. Queiroz, M. S., Souza, A. D., Abreu, E. S. V., Gomes, N. T., Neto, O. A. A., Aplicação do Processo Eletrolítico ao Tratamento de Água de Produção. CENPES-DITER-SEBIO, RT, Rio de Janeiro, Brazil (1996). (In Portuguese).

25. Ramalho, A. M. Z., Huitle, C. A. Silva, D. R., Application of electrochemical technology for removing petroleum hydrocarbons from produced water using a DSA-type anode at different flow rates. Fuel, 89, p. 531-534 (2010).

26. Rebhun, M. and Lurie, M., Control of organic matter by coagulation and floc separation. Water Sci. Technol., 27, p. 1-20 (1993).

27. Rubach, S., and Saur, I. F., Onshore testing of produced water by electroflocculation. Filtrat. Sep., 34, p. 877-882 (1997).

28. Thomas, J. E., Fundamentals of Petroleum Engineering. 2nd Ed. Rio de Janeiro, Interciência (2004).

29. Tir, M. and Mostefa, N., Optimization of oil removal from oily wastewater by electrocoagulation using response surface method. J. Hazard. Mat., 158, p. 107-115 (2008).

30. Vasudevan, S., Lakshmi, J. Sozhan, G., Effects of alternating and direct current in electrocoagulation process on the removal of cadmium from water. J. Hazard. Mat., 192, p. 26-34 (2011).

31. Yang, C., Electrochemical coagulation for oily water demulsification. Sep. Purif. Tech., 54, p. 388-395 (2007).

32. Yousuf, M., Mollah, A., Schennach, R., Parga, J. R., Cocke, D. L., and Electrocoagulation (EC): Science and applications. J. Hazard. Mat., 84, p. 29-41 (2001).

The Influence of Asphaltenes Subfractions on the Stability of Crude oil Model Emulsions

Siller O. Honse; Claudia R. E. Mansur; Elizabete F. Lucas

Instituto de Macromoléculas, Universidade Federal do Rio de Janeiro, Av. Horácio Macedo, 2030, Ilha do Fundão, 21941-598 Rio de Janeiro-RJ, Brazil

ABSTRACT

Crude oil is produced as water-in-oil emulsion, and asphaltenes have been considered the main responsible by their stabilization. The aim of this work was to evaluate the influence of the asphaltenes subfractions on the stability of petroleum model emulsions and on the efficiency of demulsifiers. Model water-in-oil emulsions were prepared: aqueous phase of brine and oil phase of asphaltenes in toluene. Different asphaltenes fractions were used. The emulsions' stability was assessed by the bottle test, with and without adding demulsifier. The results show that a sample of asphaltenes with broad polarity distribution promotes greater emulsion stability than a sample with narrow distribution and intermediate polarity. Besides this, the efficiency of demulsifiers in separating the emulsions is directly related to the original stability of the emulsion. Measurements of the interfacial tension revealed the efficiency of displacement of the asphaltenes by the demulsifiers, which occurred more efficiently for the emulsions containing asphaltenes fractions with narrow distribution and intermediate polarity.

INTRODUCTION

During production of crude oil, a large amount of water is also produced, coming from the reservoir itself and/or return of the water injected to enhance the oil recovery. In the presence of sufficient shear force when the oil and water are produced, stable emulsions can be formed at practically all steps of production and processing, such as in reservoirs, risers, treatment installations, pipelines and refineries. Once formed, the water-in-oil (W/O) emulsion can: hamper the petroleum treatment; cause changes in gas-oil separation units; affect the size of the pumping systems; and produce scaling and corrosion of equipments. W/O emulsions must be treated to remove the associated water and inorganic salts, to obtain oil with proper specifications for transport, storage and exportation and to reduce the corrosion and contamination of catalyzers at processing plants.[1, 2]

W/O emulsions are stabilized by emulsifiers (surfactants), which tend to migrate and concentrate at the W/O interface, forming a film that reduces the interfacial tension between the phases, promoting the dispersion of water droplets in the continuous phase and inhibiting their coalescence. Some natural emulsifiers are present in crude oil, such as asphaltenes, resins and organic acids and bases. Others are injected for some type of operation/treatment, such as wax deposition inhibitors, asphaltenes stabilizers and anti-corrosive agents. Fine solids can also promote mechanical stabilization of emulsions.[3-6]

The literature[7-9] shows that the molecular aggregates formed by asphaltenes, more so than the molecules in free form, help to stabilize W/O emulsions by forming a film or barrier at the interfaces.

Asphaltenes have macromolecular characteristics and are present in the heaviest fraction of crude oil. The asphaltenes fractions are also the most polar in the oil. The structure of asphaltenes is formed by aromatic polycondensate nuclei linked to the cyclical and aliphatic chains, containing heteroatoms like oxygen, nitrogen and sulfur along with metals such as iron, vanadium and nickel. Their exact structure is unknown due to the variety and complexity of their chemical structure. Several structural models have been proposed, such as archipelago and island.[10] during crude oil refining in a fractioning column, the asphaltenes are not distilled and remain solidified with the resins, receiving the name asphaltic residue. The asphaltenes are separated from the resins by the addition of an apolar (paraffinic) solvent such as n-pentane or n-heptane, forming precipitates, and are dissolved in aromatic compounds such as toluene.[11, 12]

Emulsion destabilization can be achieved by different methods.[13-15] Crude oil can be demulsified by adding chemical compounds (normally at concentrations of 10 to 1000 ppm) to improve the separation rate of the W/O emulsion. These additives cause thinning of the interfacial film, allowing the droplets to coalesce more easily, thus allowing the phases to separate.[5,16,17] These chemical compounds have nonionic character, with relatively high molar mass (normally above 3,000 Da) and have one part that is hydrophilic and another hydrophobic.[18] The hydrophilic part includes the oxyethylene, hydroxyl, carboxyl or amine groups, while the hydrophobic parts are composed of alkyl, alkylphenol or oxypropylene groups.[19,20] Among the commercial demulsifiers are ethoxylated phenol-formaldehyde resins and poly(ethylene oxide)-poly(propylene oxide) (PEO-PPO) block copolymers.[21-24]

Among the properties desired of demulsifiers are high adsorption speed at the water-oil interface, displacement of the natural emulsifiers that stabilize the emulsions and formation of thin and fragile films at the water-oil interface, facilitating coalescence of the droplets.[16, 25, and 26]

The aim of this work was to study the influence of crude oil asphaltenes subfractions, obtained and characterized previously, [27] on the stability of water-in-oil (W/O) emulsions. The measurements were performed by the bottle test, with and without the addition of a demulsifier.

EXPERIMENTAL

Materials

The asphaltic residue (ASPR) was supplied by Petrobras (Rio de Janeiro, Brazil). The asphaltenes fractions and subfractions obtained from this residue and utilized in this work were: fraction C5, subfraction C5-C6, subfraction C8-C9 and fraction C10. Such (sub) fractions were obtained in a previous work as following.[27] the asphaltic residue and n-pentane (proportion of 15g: 1L) were placed under stirring for 24 h. The insoluble fraction was placed in a Soxhlet extractor with n-pentane (proportion of 1g: 45mL). This extraction step continued until the paraffinic solvent appeared clear in the extractor. The solvent in the extractor was then replaced with dry toluene (proportion of 1 g of precipitate: 35 mL of solvent) and the extraction process was repeated until this new solvent appeared clear. The dissolved asphaltenes (fraction C5) were recovered after evaporation of the toluene in a rotary evaporator and dring for 3 days in a chapel to evaporate the residual solvent. Successive extractions were performed from the C5 asphaltenes to obtain different subfractions, separated by the difference in solubility in various solvents (n-hexane, n-heptane, n-octane, n-nonane and n-decane). The subfraction C5-C6 corresponds to that extracted by solubilization in n-hexane, starting from the C5 asphaltenes. The remaining precipitate was then subjected to another extraction, this time with n-heptane, to obtain the subfraction C6-C7 dissolved in the n-heptane. The remaining precipitate was again submitted to extraction, this time using n-octane, to obtain the subfraction C7-C8. Next, the undissolved solid was placed in n-nonane in the extractor to obtain the subfraction C8-C9. Finally, the remaining precipitate was subjected to extraction with n-decane to obtain the dissolved C9-C10 subfraction, leaving a precipitate which was called C10 asphaltenes. All the dissolved subfractions were dried for around 3 days and weighed to calculate the yield. Figure 1 shows the fractioning carried out and the yield of the asphaltenes (sub) fractions.

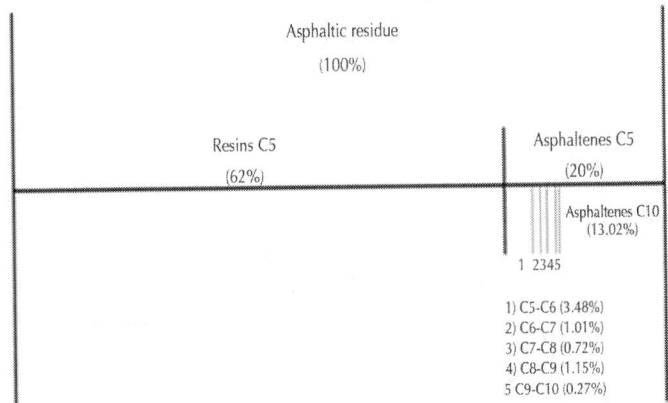

Figure 1: Yield of the asphaltene fractions.

Toluene, acquired from Vetec Química Fina (Rio de Janeiro, Brazil), was distilled and dried in alumina. HPLC-grade toluene was supplied by Tedia Brasil (Rio de Janeiro, Brazil). Sodium chloride and calcium chloride were also acquired from Vetec Química Fina.

The poly (ethylene oxide)-poly (propylene oxide) (PEO-PPO) block copolymers (branched, B; and linear, L) were donated by Dow Química Ltda. (São Paulo, Brazil). The characterization data are in

Table 1: Preparation of the model water-in-oil emulsions

Additives	$\overline{M}_n$ [a]	$\overline{M}_w$ [a]	$\overline{M}_w/\overline{M}_n$ [a]	EO/PO ratio[b]	Molecular structure
Copolymer B	11600	12000	1.03	0.19	$CH_2(PO)_{58}(EO)_{11}OH$ | $CH(PO)_{58}(EO)_{11}OH$ | $CH_2(PO)_{58}(EO)_{11}OH$
Copolymer L	3000	4100	1.37	0.51	$CH_3(PO)_{37}(EO)_{19}OH$

Ethylene oxide/propylene oxide; (a) by size exclusion chromatography (SEC); (b) by hydrogen nuclear magnetic resonance (1H-NMR).

Preparation of the Model Water-in-Oil Emulsions

First 500 mL of dispersions were prepared containing 0.25% (m/v) of one of the following samples dissolved in dry toluene: asphaltic residue, asphaltene fraction C5 (extracted with n-pentane), subfraction C5-C6, subfraction C8-C9 and fraction C10.

The model emulsions containing these dispersions as the oil phase were prepared so as to contain 30.0% synthetic brine at a concentration of 55,000 ppm of salts (NaCl:CaCl$_2$ ratio of 10:1). To prepare each emulsion, 70.0 mL of this dispersion was placed in a 250-mL beaker and submitted to shearing in a Polytron PT 3100D homogenizer under stirring of 8,000 rpm with slow addition of 30.0 mL of the brine. Then the system was kept under stirring for an additional 3 min, at room temperature.

Gravitational Water-Oil Separation Measurement by the Bottle Test

The gravitational separation of the water from the oil was measured by the bottle test. The tests were conducted initially to assess the stability of the model emulsions without the addition of a demulsifier (blank test). Afterwards, 100 ppm solutions of the PEO-PPO block copolymers in toluene (at 40% m/v) were added to the model emulsions to perform the tests with the demulsifiers added. The test procedure was described in a previous publication.[29]

The efficiency of gravitational separation of each formulation used in these tests was calculated by applying equation 1

$$EF_{WO} = (V_{WS}/V_{WT}) \times 100$$

(1)

In equation 1, EF_{WO} is the efficiency of gravitational separation of water and oil, in % by volume; V_{WS} is the volume of water separated during the test and V_{WT} is the volume of total water inside the test tube, both in mL. All measurements were taken in triplicate.

Determination of the size distribution of the saltwater droplets in the model emulsions

An Axiovert 40 MAT optical microscope was used to study the size of the saltwater droplets in the emulsions. One minute after preparing each emulsion, a small aliquot was diluted in spindle oil, placed on a slide and examined under the microscope. This was done at room temperature, with a total of 140 to 200 droplets, to obtain their average diameter by using AxioVision 4.4 imaging software.

Determination of the Interfacial Tension between the Brine and Asphaltene Dispersions

The interface tension measurements were determined by the Du Noüy ring method using a Krüss K10ST digital tensiometer, at 25 °C. All the measurements were performed in triplicate, and only the values with variation less than 1 mN m^{-1} were considered.

RESULTS AND DISCUSSION

Water-Oil Gravitational Separation Tests

The bottle test technique was used to assess the water-oil gravitational separation of the model emulsions, composed of brine as the aqueous phase and asphaltenes dispersed in toluene as the oil phase. The types of asphaltenes employed were: asphaltic residue (ASPR), fraction C5, subfraction C5-C6, subfraction C8-C9 and fraction C10. The other subfractions were not used in these tests because of the low yield during their extraction.[2]

The concentration of ASPR and the asphaltene fractions/subfractions in the oil phase was in all cases 0.25% m/v, because this was the lowest concentration tested that produced sufficiently stable emulsions to conduct the study.

All the tests were performed in triplicate and in the presence or absence of the demulsifiers (linear and branched PEO-PPO block copolymers) at a concentration of 100 ppm.

Tests without the demulsifier additives

Table 2: summarizes the maximum separation values attained for each type of emulsion

Model emulsion	Maximum separation efficiency / % (min)*
ASPR	100 (35)
Fraction C5	18 (65)
Subfraction C5-C6	100 (10)
Subfraction C8-C9	60 (25)
Fraction C10	0

Time required to achieve maximum efficiency.

The synthetic emulsions prepared without any demulsifier (blank tests) containing the ASPR dispersions as the oil phase were highly unstable. The presence of resins in the composition, which are less polar than asphaltenes, reduced the emulsifying action of this mixture of resins/asphaltenes, permitting faster coalescence of the water droplets and total separation of the aqueous phase.

The demulsification test with the oil phase composed of the C5 asphaltenes fraction presented only a small phase separation (approximately 20%), which was expected since the asphaltenes separated from resins tend to leave the solution and migrate to the interface, where they act to stabilize the emulsion.

In the blank test performed with fraction C10, the emulsion was very stable, without any phase separation during the test period (65 min). As shown in a previous work, [26] this fraction has higher polarity, so a more rigid interfacial film is generated, impeding the coalescence of the dispersed phase.

The first subfraction isolated from the asphaltenes, called subfraction C5-C6, is constituted of a portion of the C5 asphaltenes with lower polarity, leading to a less stable emulsion. Subfraction C8-C9 has higher polarity than subfraction C5-C6 and the result of the blank test showed, as expected, a more stable emulsion, with no total phase separation observed during the test.

Comparing all the results, the instability of the model water-in-oil emulsions was in the following decreasing order in terms of type of asphaltenes added:

C5-C6 > ASPR > C8-C9 > C5 > C10

It is interesting to note that the most unstable system was that containing the subfraction C5-C6. The system containing ASPR, despite the presence of the resins, was more stable than that containing C5-C6. This can be attributed to the fact that the ASPR sample contains a much wider distribution of molecular structures than the subfraction C5-C6. In other words, the ASPR sample contains the same molecules as the subfraction C5-C6, and also more apolar and more polar molecules. The presence of more polar molecules makes the system more stable, despite the presence of less polar molecules (resins) in a much higher quantity, which tend to stabilize the asphaltenes in the oil phase. The stability result leads to the conclusion that the asphaltenes fraction with narrower distribution and intermediate polarity, in this case subfraction C5-C6, has less capacity to stabilize emulsions than does a mixture of various components with varied polarities, whose more polar molecules can migrate to the interface.

Similar behavior was observed when comparing the stability results obtained for the C5 asphaltenes fraction, with wider distribution of structure types with distinct polarities, and those for the subfraction C8-C9, with narrower distribution. The subfraction C8-C9, which belongs to half of asphaltenes fraction C5 (see Figure 1) of which has lower polarity, leads to more unstable emulsions than the C5 fraction. Therefore, the difference in stability was much more accentuated: 60% separation efficiency for C8-C9 *versus* 18% for C5.

Furthermore, the influence of the C5 and C10 fractions on the emulsions' stability is very significant: the separation efficiency of 18% for fraction C5 and 0% for C10.

Tests in the Presence of the Demulsifier Additives

The linear (L) and branched (B) PEO-PPO block copolymers used here were employed in a previous work[28] as demulsifiers, and it was observed that the branched copolymer was more efficient in breaking the synthetic water-in-oil emulsions tested. This efficiency was associated with: (*i*) their branched structure, where the EO and PO groups are more distributed in the chains, thus facilitating their dispersion between the phases of the W/O emulsion, and (*ii*) their average molar mass, since too long chains cannot diffuse through the oil phase because they form agglomerates and too short polymer chains diffuse too slowly, so requiring a molar mass near to an optimal value.[30]

In this work it was also observed in the gravitational separation tests that the branched PEO-PPO block copolymer tended to be more efficient.

The separation efficiency percentages are presented in Table 3. In all the systems there was 100% separation efficiency, with only the time required to reach this result varying. This also reflects the stability of the emulsion. This complete efficiency result was to a certain extent expected, since model systems are more easily separated than petroleum emulsions.[6] The initial stability of the model emulsion is reflected in the performance of the demulsifier: the more unstable emulsions (subfractions C5-C6 and C8-C9) were broken down faster by the demulsifiers (5 and 10 min), while it took longer (15, 20 and 30 min) longer for the demulsifiers to break down the more stable emulsions (fractions C5 and C10).

Table 3: Percentage of efficiency of phase separation

	Maximum separation efficiency/% (min)*		
Model emulsion	Without additive	PEO-PPO B	PEO-PPO L
ASPR	100 (35)	100 (5)	100 (10)
Fraction C5	18 (65)	100 (15)	100 (30)
Subfraction C5-C6	100 (10)	100 (5)	100 (10)

| Subfraction C8-C9 | 60 (25) | 100 (5) | 100 (5) |
| Fraction C10 | 0 | 100 (20) | 100 (20) |

*Time required to achieve maximum efficiency

Determination of the Size Of The Water Droplets Dispersed In The Model Emulsions

Table 4 shows the average sizes, the standard deviation and the size range detected. The results show that the mean diameters of the water droplets, as well as the size distributions, were very similar in the four cases, indicating there is no influence of the water droplets' size and size distribution on the stability differences of the emulsions. This means that the differences in stability observed here really came from the type of asphaltenes molecules added to the systems.

Table 4: Mean diameter and standard deviation for different asphaltene fractions and subfractions

	Asphaltenes C5	Subfraction C5-C6	Subfraction C8-C9	Asphaltenes CIO
Diameter/μm	12.28	18.91	15.76	12.66
Standard deviation	5.98	14.62	12.35	5.51
Detected range/μm	0-40	0-120	0-130	2-32

It is important to highlight that the continuous phase in model emulsions is not as complex and viscous as that in real petroleum emulsions, so micrometer emulsions were formed and detected by optical microscopy. Literature shows that in crude oil very small drops, with sizes of some nanometers, have been detected. In this case, calorimetric measurements can be used.[31]

Determination of the Saltwater/Asphaltene Dispersion Interfacial Tensions

For the interfacial tension measurements (Table 5), each sample was placed in a cuvette and left at rest for 1 h so as to allow similar measurement conditions for all the samples. After adding the surfactant, its molecules should migrate to the W/O interface, displacing the asphaltenes molecules and promoting demulsification.

Table 5: Interfacial tension of brine/asphaltene dispersions

Asphaltenes	Interfacial tension/(mN m⁻¹)		
	Without surfactant	With surfactant B	With surfactant L
Blank*	30.5 ± 1.1	14.8 ± 0.7	11.4 ± 0.4
ASPR	20.5 ± 0.4	15.6 ± 0.4	11.2 ± 0.2
Fraction C5	21.4±0.1	16.3 ± 0.3	11.1 ±0.6
Subfraction C5-C6	26.9 ± 0.6	13.5 ± 0.8	11.3 ± 0.0
Subfraction C8-C9	21.9 ± 0.4	15.6 ± 0.5	11.0 ± 0.4
Fraction C10	20.3 ± 0.1	16.0 ± 0.4	10.3 ± 0.1

*Brine/toluene.

The blank system was composed of only brine with toluene, without the addition of any asphaltenes. In this system the interfacial tension was 30.5 mN m⁻¹, a figure that declined to 14.8 and 11.4 mN m⁻¹ with the addition of the branched and linear surfactant, respectively. As expected, in both cases the samples showed the effects of the surfactant, with the molecules migrating to the interface and reducing the interfacial tension. The linear surfactant was slightly more efficient in reducing the tension, probably due to its more hydrophilic character (EO/PO ratio = 0.51) in comparison with the branched surfactant (EO/PO ratio = 0.19): The more hydrophilic the surfactant dissolved in the organic phase is, the greater is its tendency to migrate to the interface and interact with the aqueous phase.

The first column of Table 5 contains the results of the emulsions without surfactant addition. As expected, the surfactant character of the asphaltenes can be noted, since they were able to reduce the interfacial tension of the brine/toluene system. In this case, the more accentuated the polar character of the fraction is, the greater the reduction in the interfacial tension. Subfraction C5-C6 by itself presented the highest tension value, due to its less polar character and a lesser tendency to migrate to the interface, related to the greater instability in the emulsions formed with this subfraction (Table 3). Except for the emulsion containing ASPR, the correlation between the reductions of the interfacial tension, imposed by the presence of the asphaltenes, and the stability of the emulsions remained steady: the more stable the emulsion, the lower the interfacial tension value.

It is believed that the demulsification process occurs by removal of the asphaltenes from the interface, with their place taken by the surfactant additive. This surfactant forms a less rigid interfacial film than that formed by the asphaltenes, making the emulsion more fragile, i.e., allowing the water droplets to coalesce more easily.[6, 9]

The action of the branched surfactant (B) in displacing asphaltenes from the interface was greatest in the emulsion containing the least polar subfraction, C5-C6. The interfacial tension values were higher in all the other emulsions. This behavior is likely to be related to the presence of asphaltenes molecules at the interface together with the branched surfactant molecules. Therefore, the higher interfacial tension values of the systems containing the C5 and C10 fractions than in any of the other systems was probably associated with the fact that the branched surfactant was less effective in displacing these kind of asphaltenes from the interface. This hypothesis is supported by the results on maximum efficiency presented in Table 3, where it can be seen that the systems that needed the longest time to reach 100% efficiency were those containing the C5 and C10 fractions. This correlation of results leads to the conclusion that the separation efficiency is directly related to the facility of displacing the asphaltenes molecules from the interface.

For the linear surfactant (L), all the interfacial tension values were similar to each other and to that of the system without the presence of asphaltenes. It appears that after one hour all the asphaltene is removed from the interface by the linear surfactant. None of these emulsions took more than 30 min to separate completely.

Comparison of the separation efficiency results (Table 3) with those on interfacial tension shows that lowest tension values corresponded to the lowest separation efficiencies. In other words, the systems containing the linear surfactant presented lower tension values and also less efficient separation.

CONCLUSIONS

The stability of the emulsions, as expected, was the greatest in the presence of more polar asphaltenes fractions. However, this work presents the first verification that a sample of asphaltenes with wide polarity distribution promotes greater stability than a sample with much narrower distribution and intermediate polarity. This means that even the presence of a large quantity of less polar molecules is not sufficient to stabilize the more polar molecules in the dispersion, causing these molecules to migrate to the emulsion's interface.

The efficiency of demulsifiers in separating emulsions is related to the original stability of the emulsion, i.e., more unstable emulsions are broken down by demulsifiers more quickly. In this work, all the emulsions containing demulsifiers were completely separated, due to the greater facility of separating the model emulsions (which are composed of brine and a dispersion of asphaltenes in toluene) than petroleum emulsions. The greater efficiency of the branched surfactant over the linear one was confirmed.

The size and size distribution of the water droplets in the model emulsions were similar, indicating the effect of the type of asphaltenes molecules on the stability of emulsions.

The interfacial tension measurements revealed the efficiency of the demulsifier surfactants in displacing the asphaltenes, which occurred more efficiently for the emulsions containing fractions with narrow distribution and intermediate polarity.

ACKNOWLEDGMENTS

We thank CNPq, CAPES, FAPERJ and Petrobras for financial support and Professor Marcio Nele (School of Chemistry of Rio de Janeiro Federal University) for the use of the inverted optical microscope.

REFERENCES

1. Alboudwarej, H.; Muhammad, A.; Shahrakl, A.; Dubey, S.; Vreenegoor, L.; Saleh, J.; *SPE Prod. Oper.* 2007, *22*, 285.

2. Kokal, S.; Al-Ghamdi, A.; Meeranpillai, N. S.; *SPE Proj. Fac. Const.* 2007, *2*, 1.http://dx.doi.org/10.2118/102856-PA accessed in June 2012.

3. Midttun, Ø.; Kallevik, H.; Sjöblom, J.; Kvalheim, O. M.; *J. Colloid Interface Sci.* 2000, *227*, 262.

4. Sjoblom, J.; Hemmingsen, P. V.; Kallevik, H. In *Asphaltenes, Heavy Oils, and Petroleomics*; Mullins, O. C.; Sheu, E. Y.; Hammami, A.; Marshall, A. G., eds.; Springer: New York, 2007, ch. 21.

5. Kelland, M. A.; *Production Chemicals for the Oil and Gas Industry*, CRC Press: New York, 2009.

6. Ramalho, J. B. V. S.; Lechuga, F. C.; Lucas, E. F.; *Quim. Nova* 2010, *33*, 1664.

7. Mclean, J. D.; Kilpatrick, P. K.; *J. Colloid Interface Sci.* 1997, *196*, 23.

8. Bauget, F.; Langevin, D.; Lenormand, R.; *J. Colloid Interface Sci.* 2001, *239*, 501.

9. Spiecker, P. M.; Gawrys, K. L.; Trail, C. B.; Kilpatrick, P. K.; *Colloids Surf., A* 2003, *220*,9.

10. Mullis, O. C.; Sabbah, H.; Eyssautier, J.; Pomerantz, A. E.; Barré, L.; Andrews, A. B.; Ruiz-Morales, Y.; Mostowfi, F.; McFarlane, R.; Goual, L.; Lepkowicz, R.; Copper, T.; Orbulescu, J.; Leblanc, R. M.; Edwards, J.; Zare, R. N.; *Energy Fuels* 2012, *26*, 3936.

11. Kokal, S.; Aramco, S.; *Crude-Oil Emulsions: A State-of-the-Art Review*. SPE Annual Technical Conference and Exhibition: San Antonio, Texas, 2002. http://dx.doi.org/10.2118/77497-MS accessed in June 2012.

12. Wang, X.; Alvarado, V.; *Effect of Salinity and pH on Pickering Emulsion Stability*, SPE Annual Technical Conference and Exhibition: Denver, Colorado, 2008. http://dx.doi.org/10.2118/115941-MS accessed in June 2012.

13. Figueiredo, E. N.; Arêas, J. A. G.; Arêas, E. P. G.; *J. Braz. Chem. Soc.* 2008, *19*, 1336.

14. Oliveira, A. A. S.; Teixeira, I. F.; Ribeiro, L. P.; Tristão, J. C.; Dias, A.; Lago, R. M.; *J. Braz. Chem. Soc.*2010, *21*, 2184.

15. Yuan, S, Tong, M.; Mu, G.; *J. Hazard. Mater.* 2011, *192*, 1882.

16. Kim, Y. H.; Wasan, D. T.; *Ind. Eng. Chem. Res.* 1996, 35, 1141.

17. Lucas, E. F.; Mansur, C. R. E.; Spinelli, L.; Queirós, Y. G. C.; *Pure Appl. Chem.* 2009, *81*, 473.

18. Peña, A. A.; Hirasaki, G. J.; Miller, C. A.; *Ind. Eng. Chem. Res.* 2005, *44*, 1139.

19. Wu, J.; Xu, Y.; Dabros, T.; Hamza, H.; *Energy Fuels* 2003, *17*, 1554.

20. Zhang, Z.; Xu, G. Y.; Wang, F.; Dong, S. L.; Li, Y. M.; *J. Colloid Interface Sci.* 2004, *277*, 464.

21. Mansur, C. R. E.; Oliveira, C. M. F.; González, G.; Lucas, E. F.; *J. Appl. Polym. Sci.* 1997, 66, 1767.

22. Mansur, C. R. E.; Spinelli, L. S.; González, G.; Lucas, E. F.; *Colloids Surf., A* 1999, *149*, 291.

23. Mansur, C. R. E.; Barboza, S. P.; González, G.; Lucas, E. F.; *J. Colloid Interface Sci.* 2004, *271*, 232

24. Mansur, C. R. E.; Pires, R. V.; Gonzalez, G.; Lucas, E. F.; *Langmuir* 2005, *21*, 2696.

25. Szabó, G. H.; Masliyah, J. H.; Elliot, J. A. W.; Yarranton, H. W.; Czarnecki, J.; *J. Colloid Interface Sci.* 2005,*283*, 5

26. Dicharry, C.; Arla, D.; Singuin, A.; Garcia, A.; Bouriat, P.; *J. Colloid Interface Sci.* 2006, *297*, 785.

27. Honse, S. O.; Ferreira, S. R.; Mansur, C. R. E.; González G.; Lucas, E. F.; *Quim. Nova* 2012, *35*, 1991.

28. Silva, P. R. S.; Mauro, A. C.; Mansur, C. R. E.; *J. Appl. Polym. Sci.* 2009, *113*, 392.

29. Pacheco, V. F.; Spinelli, L. S.; Lucas, E. F.; Mansur. C. R. E.; *Energy Fuels* 2011, *25*, 1659

30. **Álvarez**, F; Flores, E. A.; Castro, L. V.; Hernández, J. G.; López, A.; Vasquez, F.; *Energy Fuels* 2011, *25*, 562.

31. Díaz-Ponce, J. A.; Flores, E. A.; Lopez-Ortega, A; Hernández-Cortez, J.; Estrada, A.; Castro, L. V.; Vasquez, F.; *J. Therm. Anal. Calorim.* 2010, *102*, 899.

Citations

CHAPTER 1

K. K. Salam, A. O. Alade, A. O. Arinkoola, and A. Opawale, "Improving the Demulsification Process of Heavy Crude Oil Emulsion through Blending with Diluent," Journal of Petroleum Engineering, vol. 2013, Article ID 793101, 6 pages, 2013. doi:10.1155/2013/793101.

CHAPTER 2

Pingting Liu, Zhiyu Huang, Hao Deng, Rongsha Wang, and Shuixiang Xie, "Synthesis and Performance Evaluation of a New Deoiling Agent for Treatment of Waste Oil-Based Drilling Fluids," The Scientific World Journal, vol. 2014, Article ID 852503, 9 pages, 2014. doi:10.1155/2014/852503.

CHAPTER 3

Mohammed Yahaya Khan, Z. A. Abdul Karim, Ftwi Yohaness Hagos, A. Rashid A. Aziz, and Isa M. Tan, "Current Trends in Water-in-Diesel Emulsion as a Fuel," The Scientific World Journal, vol. 2014, Article ID 527472, 15 pages, 2014. doi:10.1155/2014/527472.

CHAPTER 4

Ramesh P Babu, Kevin O'Connor, and Ramakrishna Seeram, Current Progress on Bio-Based Polymers and their Future Trends, doi:10.1186/2194-0517-2-8.

CHAPTER 5

M. Keshavarz Moraveji, E. Mohsenzadeh, M. Ebrahimi Fakhari, and R. Davarnejad, Hydrodynamics and Oxygen Mass Transfer Characteristics of Petroleum Based Micro-Emulsions in a Packed Bed Split-Cylinder Airlift Reactor, ISSN 0104-6632.

CHAPTER 6

Fabiane G. Antes; Juliana S. F. Pereira; Liange O. Diehl; Letícia S. F. Pereira; Paula Boeckl; Regina C. L. Guimarães; Ricardo A. Guarnieri; Bianca M. S. Ferreira; Maria de Fatima P. dos Santos; Edson Luiz Foletto; Erico M. M. Flores, Sediment Removal from Crude Oil Emulsion Using Microwave Radiation, doi.org/10.5935/0103-5053.20130163.

CHAPTER 7

A. A. Cerqueira, P. S. A. Souza; M. R. C. Marques, Effects of direct and alternating current on the treatment of oily water in an electroflocculation process, http://dx.doi.org/10.1590/0104-6632.20140313s00002363.

CHAPTER 8

Siller O. Honse; Claudia R. E. Mansur; Elizabete F. Lucas, The influence of asphaltenes subfractions on the stability of crude oil model emulsions, doi.org/10.1590/S0103-50532013005000002.

Index